ACKNOWLEDGMENTS

THE AUTHORS take pleasure in acknowledging their great indebtednesses: to Professor Edward L. Bowles, of the Department of Electrical Engineering at Massachusetts Institute of Technology, for a minute reading of the chapters dealing with electricity; to Professor Joseph C. Riley, of the Institute's Department of Mechanical Engineering, for a similar examination of the chapters relating to steam; to Mr. Edward Hungerford, of the New York Central Railroad, for his helpful criticisms of the four chapters on the development of the locomotive and railroads; to Mr. F. X. Milholland, of the Baltimore & Ohio Railroad, for his courtesy in supplying photographs from the invaluable archives of that railroad; and to Mr. Carl H. Claudy for a quizzical study of the entire manuscript, and many helpful suggestions.

The authors' secretariat displayed such unusual ability to transform cuneiform inscriptions into legible typescript that proper acknowledgment must be made in other ways and other places.

BEHEMOTH

THE STORY OF POWER

BY ERIC HODGINS AND
F. ALEXANDER MAGOUN

"I sell here, Sir, what all the
world desires to have: *Power!*"
—MATTHEW BOULTON TO JAMES BOSWELL

MCM XXXII
DOUBLEDAY, DORAN
& COMPANY, INC.
GARDEN CITY, N. Y.

PRINTED AT THE *Country Life Press,* GARDEN CITY, N. Y., U. S. A.

FIRST EDITION

DISCLAIMER

THIS is a very unlearned book, as the persevering reader will discover for himself in a few pages. The authors have little idea of what is wrong with "our machine civilization" and have labored to avoid giving any accidental impression to the contrary. They are in the odd emotional state that the economic pronouncements of, say, Dr. Julius Klein, fill them with longing for a world of spinning wheels, parched corn, and goat's milk—yet the medievalism of Mr. Ralph Adams Cram is just as likely to drive them, sobbing, into the arms of Mr. Charles M. Schwab. The horns of a dilemma, as a more gifted pen has set down, are sounding outside their door. Their book, then, defends no one and indicts no one. Its purpose is purely narrative.

It has been suggested that references to a "Conservative Party" which occur throughout the book may confuse occasional readers—particularly if they belong to it. These references have nothing to do with Lord Randolph Churchill, Sir Robert Peel, or Calvin Coolidge. The authors have used the term merely as a catch-all for those gentlemen who, whenever they have been faced with something new in the physical world, have always hastened to denounce it for its supposed effect upon manners or morals, or its lack of sanction by the Old Testament.

Now and again the reader with a technical background may be puzzled by over-simplifications or omissions. The former have been chosen in preference to bewildering pages of technical explanation; the latter, when not accidental, are explained by the necessity of hewing closely to the story of Power, letting theories fall where they would. No book on electricity could be excused for so cavalier a treatment of James Clerk Maxwell or Charles Proteus Steinmetz as this one permits itself; but the authors submit another reminder that this is merely a narrative of adventure, in an intensely physical, non-abstract world.

E. H.
F. A. M.

New York and Cambridge
March, 1932

CONTENTS

ILLUSTRATIONS

BEHEMOTH

CHAPTER I

Muscles for Machinery

Whatever may be (and there are a good many guesses) the oldest civilization in man's history, there is no question that it had at its disposal a unique motor. Since civilization might be roughly described as an effort on the part of men to leave things better organized than they found them, some sort of power has been a constant necessity for the performance of work.

In the earliest days the world's first engine operated with a surprising efficiency. It was not really one motor but a combination of over three hundred engines, large and small, gathered together under one system of automatic control. None of these had a flywheel, yet each always avoided dead centers. Some of the larger ones contained as many as six hundred thousand small cylinders, joined together in compound. The boiler was capable of burning fuel at a very low temperature. Whereas it now takes ten hours to get up steam on a ship and three or four on a locomotive, this early motor was almost instantly ready for action. It worked at a constant pressure and had an elaborate alarm system to warn the operator when fuel ran low. It was built exclusively on the lever principle and contained about two hundred and thirty bearings all automatically lubricated. As would be expected of so complex a mechanism it got out of order now and then, yet its length of service was surprisingly good. Twenty years is a good life for a modern steam engine but this earliest motor would usually run for more than sixty. And whereas a modern steam engine has an efficiency (a word to describe the ratio between the energy put into an engine and the energy it gives out) of only about 15 per cent, a gasoline engine of 20 to 25 per cent, and a Diesel, or crude-oil-burning engine, as high as 35 per cent, this oldest of all engines would yield an efficiency of 40 per cent. For some five thousand countable years—and no one knows how many uncountable ones—it was entirely without a rival.

This engine was the human body.

The great trouble with the human body as an engine has always been that despite its great ingenuity and high efficiency its total capacity for work is quite small. In proportion to his body weight, a man can accomplish prodigies. He can, for example, shovel forty times his own weight during the course of an hour. But that has never been enough to satisfy his own ambitions—by the demands of which, his muscles are poor things. A Sicilian slaving at a sewer pump eight hours a day can develop barely one tenth of a horsepower.[1] For shorter periods a man in the best physical condition—let us say a member of a college crew straining at the tholepins of a racing shell—can develop a horsepower in the last desperate seconds of a sprint. But beyond this the action of the human heart places a strict limit on the rate of work that a human being can attain. A man laboring at a crank might exert a continuous force of fifteen to eighteen pounds on the handle and could turn it twenty-six to thirty revolutions per minute. Accordingly he might produce as many as 1,296,000 foot-pounds of work in the course of a day, whereas a one-horsepower engine would mount its total to 15,840,000 foot-pounds. There lay the disadvantage of the

[1]Let us hope that this will be the only footnote in this book. It is needed to say that the word "horsepower" was first defined and used by James Watt. It is the equivalent of 550 foot-pounds per second. A foot-pound in turn is the amount of work done when a weight of one pound is lifted one foot. A horse, not knowing any arithmetic, cannot exert a continuous horsepower for more than a few hours.

slave; to exert a horsepower, he would need a body which weighed 1,500 pounds. The steam engine will produce a horsepower for 150 pounds, and a modern aviation engine will compress a horsepower into two pounds. If a new-born baby were as powerful, it would have the strength of forty men. The strongest man in the modern world was perhaps P. J. McCarthy, a professional weight lifter, who once astounded the world of 1898 by supporting a platform loaded with 6,370 pounds of stone. But what was a prodigy for a man would have been a mean performance for a steam engine.

In a section of land there are 640 acres. An active man, in the days before power, might have worked his life long in vain to spade a fraction of it. If mechanical power had never been invented this man might have begun to spade his section in the year 1432. To-day, but not with the same shovel, his great-great-great, and then many times more great, grandchild would just be coming to the end of a five-hundred-year task. But thanks to the thoughtful gentlemen who brought mechanical power into the world, three men with a tractor and gang plows could turn up every inch of that soil within thirty-six hours.

With no sources of power save muscles, ten thousand men might have begun to dig the Panama Canal in the year that the Pilgrims landed at Plymouth. Assuming that they knew all about sanitation, they might hope to finish the mere job of digging the 232,353,000 cubic yards of earth and rock by a little after the year 2000.

The broad jump of an athlete might be twenty feet. He might travel as fast as ten miles per hour on the take-off and might arch four feet above the ground at the summit of his rise. But that is just about as long and as fast and as far as a man alone can get. Yet a man-made airplane in this modern day of power may make a jump that is five thousand miles long and twenty thousand feet high and do it at a speed well above one hundred miles per hour.

A vigorous man can carry one hundred pounds for fifteen miles in a day, but he will find himself spent and exhausted at the end. Yet a scrubby freight locomotive can rocket through the night, towing two thousand tons behind it and travel two hundred and fifty miles from dusk to dawn. And then from dawn to dusk it could do the same thing all over again with the barest pause for those meager stimulants like coal and oil and water which machinery finds sufficient to sustain its life.

The *Bremen* and the *Europa,* holders of the blue ribbon for speed on the North Atlantic, each pack one hundred thousand horsepower into boiler rooms not much larger than a triplex apartment on Park Avenue, New York. If men alone were to produce this power as they produced it on Roman triremes centuries ago, Atlantic crossings would be rather difficult. One man can exert one tenth of a horsepower for eight hours; accordingly each of these ships would find it necessary to impress into service about three million men—a population greater than that of the entire city of Philadelphia—in order to carry two thousand dinner-jacketed gentlemen and seventeen-jeweled ladies on the high seas at twenty-seven knots.

Perhaps this isn't progress; progress means improved sewage disposal to some people; to others it means community singing or the rise of a new political party. In these pages we need take neither one side nor the other. The present machine civilization of most of the world has violent partisans (usually people who make money out of it) and just as violent detractors (usually those who don't make money out of it). These latter feel that civilization is collapsing of its own complexity, and that a locomotive driver has a harder life than a galley slave. However that may be, those who dislike machines must put the blame for their existence where it belongs: nature's irremediable error in permitting water to become a gas when sufficiently heated, and man's inexcusable impertinence in sufficiently heating it.

Up to the middle of the eighteenth century—before heat and power became linked together in men's minds—the kings and generals who were the managers of the world's affairs had only muscles to aid them. The best managed but weakest muscles belonged to men. They needed reinforcement, and they got it from the muscles of horses, oxen, and asses and, upon the resourceful continents of Asia and Africa, also from the muscles of elephants, camels, and dromedaries. They were helped out by a few rudimentary machines, as will later become evident, but essentially muscles were the only power plants. Engine in himself, and stoker for all other engines, was the slave.

The work that a slave could accomplish was not dependent wholly upon such simple things as temperature and lubrication. His health and muscular development, his nervous temperament and disposition, his race and training, the food he ate, the air he breathed, the climate he worked in, the number of hours a day he worked—all these and many other things determined the amount of work that he could do. His labors were to push, or to pull, to twist, or to lift. His ordinary push or pull was twenty-six pounds, his lift was forty-four. His shovel carried six pounds and the force of his hammer blow was fifteen.

In the great days of Greece, each freeman owned on the average two slaves. Of Rome, in the days of its greatest glory, Gibbon thought there were as many slaves as freemen. One Cæcilius, a freeman in the days of Cæsar Augustus, was distinguished by the possession of 4,116 slaves. That seems an opulent number, yet by modern standards they were the equivalent of a mere four hundred horsepower—something less than is now housed every evening under the roof of any public garage. Had Cæcilius then had the advantage of knowing Shakespeare he might have paraphrased *As You Like It* and of his best slave said: "An ill favored thing, Sir, but mine only."

The slave left his monuments. He built the Great Wall of China. He built the Pyramids. He built the Roman roads and aqueducts. He was the essence of progress and the measure of wealth in pre-Christian times. But often he must have thought under the sting of an overseer's lash that he might have been better off in simpler days, when cavemen, having no use for conquered enemies, killed them outright. Already, the world's work was an insufferable burden upon the shoulders of men.

Burton Holmes-Galloway

CAPACITY, ONE OX-POWER

THIS rudimentary machine for lifting water is still used in Portugal

Cowling-Galloway

THE FACTORY SYSTEM IN EGYPT

MAN-DRIVEN crushing wheels are still the last word in power

Ewing Galloway

THE MACHINE AGE IN THE ORIENT

THREE Hindus using a cocoanut palm trunk as a sweep to lift water from an irrigation ditch

Burton Holmes-Galloway

PUMP AND ENGINE

A SIAMESE woman driving a treadmill irrigation pump in a ricefield

CHAPTER II

When Horses Were Wishes

Man, a sentimental animal, has now and again referred to the horse as "man's best friend." It would be interesting to hear the horse's version of this.

Man's best friend began by serving man as food. Wild horses were common in Europe during the Stone Age; it was during this period that the horse had the pleasure of serving not for work but for nutrition. The horse, despite his ancient origin, was domesticated long after those two favorite animals of the Old Testament, the ox and the ass. Perhaps the horse had no other gesture with which to demonstrate his friendship until, ages later in Central Asia, nomadic herdsmen accustomed him to being ridden. However that may be, horses were introduced into Egypt by the Hyksos, or shepherd kings, who came from the north and east of Syria and conquered lower Egypt in the seventeenth century before Christ. Early in his career, then, the horse was regarded as an implement of war; his services in the causes of peace, hauling carts and plows, are much more recent.

For one thing, the horse is a more delicate mechanism than the ox or the ass. *These* were the ancient draft animals; horses were used for their speed. The early methods of setting animals to draft work were extremely poor; the horse remained unshod until after the fall of Rome, and early harness was so poorly designed that if a horse pulled with any great enthusiasm his breath was cut off for his pains. Competent horseshoeing and sensible harness-making were not fully established until the twelfth century. Lacking proper harness, even a pair of oxen—by the testimony of Xenophon—could draw behind them a bare fourteen hundred pounds. Oxen went down to the sea in ships, as well: the Roman Army under Claudius Caudex was transported across the Straits of Messina to Sicily in boats supposedly propelled by paddle wheels turned by oxen.

By the late seventeenth century, while Dionysius Papin was idly toying with steam pressures, the horse had risen greatly in world importance. He was no longer solely an instrument of war; he was an important adjunct to commerce. The vehicles which he drew behind him were growing more elegant in their furnishings and more refined in their delicacy. And by the time the eighteenth century arrived, England, with her length and breadth well crisscrossed by stagecoach lines, felt that the era of perfect transportation was, thanks to the horse, fast approaching. The Holyhead Mail via Shrewsbury was timed to the speed of eleven miles per hour and the London coach for Bath, about one hundred and twenty-five miles away, would start at five o'clock on Monday, Wednesday, and Friday mornings, performing the journey in three days "if God permit." By 1742 the Oxford stage was leaving London no earlier than seven in the morning and could usually make Uxbridge by noon, arriving for the night at High Wycombe by five o'clock. It took but ten hours to travel twenty-seven miles—provided, among other things, that the traffic of the day minded the sharp polished bugle note of the guard sounded in warning to clear the road ahead for the dashing horses of the team. Stagecoach travel, with its pleasant accessories of hostlers and inns, cracking whips and flying manes, and the compounded smells of whiskey and hot leather, did not reach its zenith until the year 1823—just two years before the steam engine first applied itself to locomotion upon the Stockton & Darlington Railway.

Long before that time, however, the horse had been put to more ignominious uses—chief among which was service at the coal mines. As the demands of society grew more complex and imperious the mine owners of England, then as now lagging sadly behind the times, were hard put to it to keep their mines sufficiently unwatered to be workable. When man's muscles finally proved insufficient to the task of pumping, the horse was impressed into service. The machines to which he was attached grew more and more rigorous. As mines grew deeper the work of pumping grew more and more arduous until, toward the close of the eighteenth century, at least five hundred horses were being used in one English mine, to drive treadmills connected to chains of buckets. It could not endure, and it did not.

Yet even after the advent of steam, man continued to demonstrate his friendship for the horse by applying him to strange purposes. In 1825 a Frenchman, one of the many hunting for some light power plant with which to drive an airship, drew plans for a *ballon à chevaux*. He designed a ship 152 feet overall, carrying below it a suspended platform upon which was to be constructed a treadmill. This treadmill, in the mind of M. Genêt, the inventor, was to be geared to a propeller and the propeller driven by the activities of two proud horses whose restricted gallopings would drive it swiftly through the happy skies.

This device at least had the virtue of remaining upon paper. Not so that of Mr. C. E. Detmold, an American railroader. In the year 1829 the South Carolina Railway was resolved to supply everything necessary to steam transportation except steam. It authorized Detmold to construct a horse-driven locomotive. This was to be a simple affair—a platform mounted upon flanged wheels, in the center of which stood a tethered horse upon a treadmill. Upon benches flanking the horse, and in a position vulnerable to any acts of discourtesy which it might think of, were the passengers. For the pleased Mr. Detmold this design won a $500 prize. And at least one horse sufficiently overcame the inefficiencies of his unnatural position and coupling with the ground to drive the car at twelve miles per hour. The idea died slightly before the horse, and the South Carolina Railway then adverted to steam.

Shortly afterward, the horse was privileged to participate in the civilization of New York City. In 1825 the Erie Canal had opened and New York, with a population of some two hundred thousand people, was by way of becoming an important port. Five years later the best citizens of the city (members of the Conservative Party, as always) watched with growing displeasure the laying of strap rail tracks several inches above the pavement from Twenty-third Street north. The ambitious idea of the promoters was to lay these rails all the way to Harlem River—although what business any New Yorker might have in Harlem, Heaven alone knew. It was a year when the inhabitants of the city used the phrase "above Bleecker" to indicate a residence in the purlieus. The best citizens had only this consolation: the "New York and Harlem Railway" had been granted its franchise with the proviso that if its tracks proved an impediment they were to be removed.

Within a year the emboldened promoters were digging up the cobblestones of the Bowery, that proud street, for a distance of three quarters of a mile between Union Square and Prince Street.

The result of all these ominous preparations was that on November 14, 1832, the first street car system in the world began operation. Two horses were motive power for each car; the cars had been built by the best of local carriage builders and seated eight to ten persons. Naturally, save for their wheels, they resembled the stagecoaches from which they sprang, with the driver sitting high up at the front and handling a long brake lever—supplied for safety in the dense traffic. Springs there were, for the comfort of the passengers. Each car was extravagantly named, in a tradition now carried on by the Pullman Company.

Upon this November day the mayor, council, and "strangers of distinction" drove from City Hall to observe the high festivities of inauguration in Union Square. The mayor and his council were to ride in the first carriage, driven by one Lank O'Dell. There was to be a second car also, piloted anonymously. As a safety demonstration (there had been a good deal of worry over this) the horses were to be trotted briskly, a signal given, and the brakes strongly applied. Windows along the Bowery bulged with spectators as the proud caravel started off on its first journey. Unfortunately, through the panic of the anonymous hack driver, the first of all street car safety pageants turned out to be the first street car accident. Lank O'Dell, at the signal, yanked on the reins and dragged his brake lever up to the last notch. But the second car, thundering along at some eight miles per hour, crashed into it, the wagon tongue thrusting its length into the delicate car ahead. There were, however, no fatalities and the railway line, despite the Conservative Party (which kept it off Broadway for years), flourished. By 1837 the rails had reached Harlem and by 1839 they had extended in the other direction as far south as the City Hall. Spokesman for the objectors was former Mayor Phillip Hone, who never relaxed his stringency.

> Accidents [he said] will frequently occur, especially on Sundays when the cars return to the heart of the city overloaded with passengers who have been Sabbath-breaking in the country,

and many of whom are not in a proper condition to take care of themselves.

By 1849 the old stagecoach type of car had been superseded by something approaching the modern street car—a short, dumpy, bounding vehicle set low on its trucks but with pronounced eaves to the roof and a smoke pipe from the stove projecting raffishly in the corner. For a time steam locomotives, now ascendent, supplanted horses—whereupon the accidents always predicted did generously begin to occur and the legislature passed a law which made the horse the one legal means of conveyance in city streets. There he remained until in 1917 New York's last horse car disappeared, and the horse, casting off the many mechanical burdens which had been hung upon him through his centuries of intimacy with the horse's best friend, returned (save for an occasional fire engine) to those light wagons from which he had been so rudely summoned some two hundred years before.

CHAPTER III

The Winds and the Waters

LONG years ago, Poseidon, the brother of Zeus, became, according to one genealogy, the father of a son, Æolus. Since Poseidon ruled the seas and Æolus came to rule the winds the ancient world was well provided with sources of power long before Prometheus brought his gift of fire to the grateful ancients. But unfortunately the ancients had little idea what they might do with these boons. The total world hydraulic power is the equivalent of some 240 million horses—and a wind that is one hundred feet high and a mile wide produces, at thirty miles per hour, one hundred thousand horsepower.

The winds and the waters were used as sources of power by mankind so long ago that it is not possible to say when. All we definitely know now is that fragments of Egyptian pottery, which go back four thousand years before Christ, show ships with square sails navigated by the knowing Egyptians.

No doubt the first application of a sail to a ship was resented by the Conservative Party in Egypt on the ground that it was somehow not proper to mitigate the lot of slaves by letting nature help them; that the obvious way to propel a ship was with oars driven by the aching biceps of the working classes, and that no good would come of coddling them by letting the winds help out. How the Progressive Party answered this powerful argument it is not easy to say—yet the sail persisted. The Egyptians waxed and waned, Greece was toppled over by Rome, and Rome fell before the barbarians. Meanwhile, by the aid of this sail, the Phœnicians had sailed out of the Mediterranean to discover the Canary Islands, Mark Antony had taken Cleopatra upon voyages of relaxation, and William the Conqueror had sailed to Britain and subdued it. The sail was making a name for itself, even though for at least five thousand years it had persisted, innocent of change—a single square sheet of cloth with a yard at the top and of use only in a following wind.

But even though, in the twelfth century, the power applied to ships had changed no whit from what it was in the days of the Egyptians, the helpfulness of the wind had impressed itself upon men's minds and there came into being a new machine, the windmill. In England the first definitely known windmill was built in the year 1191. Dean Herbert put it up in his glebe lands at Bury St. Edmunds and used it to grind corn. It was built in defiance of vested authority and was quickly destroyed by the abbot. Thereupon, like all things which have the advantage of being useful and illegal, windmills multiplied rapidly. By 1346 they were sufficiently plentiful to permit his majesty Edward III to watch the progress of the Battle of Cressy from the top of one.

These early windmills were built upon posts and consisted of a large fan of four sails (made of cloth or wooden slats) the axle of which ended in pinion gearing of wood or iron. Through this the fan was made to turn a vertical axle at the bottom of which rested a great millstone two or three feet in diameter—for the principal purpose of these early mills was the grinding of grain. This was a great improvement over treadmills, yet wind as a power source had two great disadvantages: it was intermittent in its force and fickle in its direction.

Toward overcoming the first of these misfortunes there has been nothing that any engineer, ancient or modern, could do—Æolus has always reserved the right to work only when

HORIZONTAL WATER WHEEL

DIAGRAM of a primitive arrangement for making water do the work of muscles

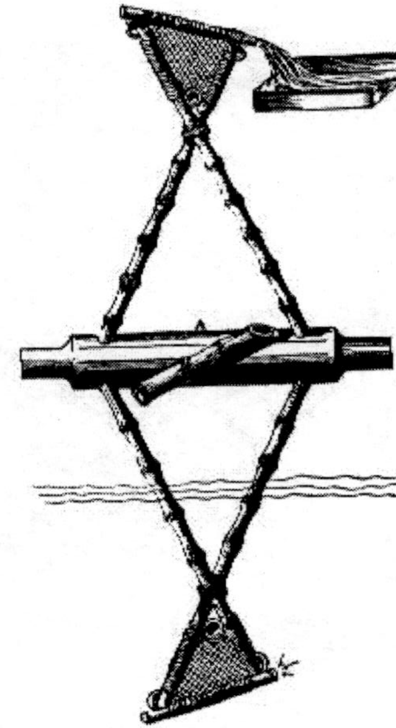

CHINESE WATER WHEEL

ONE means by which the wily Chinese raised water

Smithsonian Institution

"FLYING DUTCHMAN"

Model of one early idea for a locomotive—a horse upon a treadmill

he pleases. To overcome even the disability of direction changes early millers had trouble enough. They could think of nothing better than putting an enormous lever at the bottom of the miller's house—sufficiently long so that, when the wind changed, a crew of men could push the whole apparatus around until the sweeps were once again in the right plane. Some few centuries elapsed, and another improvement was made: a few mules were substituted for the crew of men. Mark Twain could visit England in the middle of the nineteenth century and, to his great indignation, still find this arrangement unchanged.

Because of his mercurial motive power the life of a miller was arduous and exciting. There might be days on end when the breath of wind was so light that the millstones would not turn and sacks of ungrindable grain would pile up upon the miller's floor while he sighed for the dear dead days of his ancestors when tethered animals turned the stones and modern machinery had not made it impossible for an honest man to earn an honest wage. But then the wind would spring up to a gale, and the house of the miller would sway and groan while the heavy, barbaric machinery of the mill would crash all day and night, the miller seeking desperately to catch up with his back work. Now and again a mill groaned from excess wind pressure until the whole affair collapsed on its side like an overrigged ship and the miller climbed ruefully out from among his millstones, one more victim of the machine age. As time went on, millers learned how to guard against this mishap by using the basements of their mills to store heavy sacks of grain.

The grinding of grain still remained difficult, but slowly the windmill evolved into a more sensible form. By the fifteenth century it came to be a rigid conical structure of wood, brick, or stone mounted by a revolving cap which carried the axle of the sweeps. Thus the "tower mill" slowly began to supplant the "post mill." Then in 1750 the windmill became the beneficiary of another great invention. Upon the revolving cap and on the side opposite the main sweeps a brilliant inventor named Meikle placed a smaller fan in a plane at right angles to the main sweeps. When, therefore, the wind struck the sweeps full square there was nothing to revolve the smaller fan. Now let the wind veer ever so slightly and some component of it would revolve the smaller fan which, connected with a rack and pinion, would trundle the movable mill top about until the main sails once again presented a ninety-degree front to the wind. By that simple but exciting invention the windmill became a semi-automatic machine which would keep on grinding grain so long as enough wind blew, regardless of the quarter.

The miller was now at the mercy only of the force of the wind; with fifty- or sixty-foot sails a ten-mile wind would grind five hundred pounds of grain per hour between one set of stones. A strong, consistent wind might grind four or five tons of grist for cattle during a "working day," which varied anywhere between five and twenty-four hours, depending upon fortune. The great enemy of the miller now was inconstancy. In a wind of less than eight miles per hour, less than a horsepower would be developed at his stones—probably not enough to grind. He was still likely to be idle as a dunce or busy as a maniac.

Yet not all millers were at the mercy of the winds. Some of them, in the belief that Poseidon was a more merciful god than Æolus, ground by water, not by air.

As a matter of fact the palm for antiquity goes to millers by water, not to millers by air. Water power was used for grinding grain long before the birth of Christ—once again by those canny folk, the Egyptians, who likewise had irrigation problems upon their hands and resorted to one means or another to make water lift itself to a higher level. It is quite possible that Mithradates, King of Pontus, found time in his busy life of matricide, conquests, and poisonings to own a water mill. But the earliest exact record seems to fix the year 85 B. C. as a definite date in the history of water power, for it was then that Antipater of Thessalonica is credited with having put a stream to work.

The first water wheels used to drive corn mills were horizontal—that is, they revolved in the same axis as a subway turnstile. And their function was somewhat the same—to permit nothing to pass without extracting a toll. The toll of the water wheel was energy, but as can readily be seen from the illustration facing page 14 these mills did their work with a maximum of splashing and a minimum of efficiency. Nevertheless many an ox breathed easier when the first one began the work of relieving the overwrought muscles of the ancient world.

Horizontal wheels were favored by the Norsemen, who introduced them into Ireland (there was a fight about it, no doubt), in the third century. They had two great advantages—they could be directly connected to millstones without bevel gearing, and they did not make such conspicuous targets for the enemy during the almost perpetual states of war which an earlier world enjoyed.

Yet despite these advantages the vertical type of wheel prospered the more. It had perhaps had an even earlier start, for the Chinese (who seem to have invented everything before the western world had heard of it, with the possible exception of the Model A Ford) had long used a peculiar variation of it made of their prime structural material, bamboo. By it they not only made a wheel revolve but forced it to carry small amounts of water from the bottom to the top of its circumference. With this honorable ancestry the vertical wheel grew and prospered. The Roman Vitruvius, near the year 16 B. C., described one that, against its natural inclination, was made to drive grinding stones by means of gear wheels which transferred its motion from a horizontal to a vertical axle—perhaps the first known instance when gear wheels were used in the transmission of power.

One type of vertical wheel found its earliest use in the Roman empire during the days of the first Emperor Justinian. The first idea of the Romans (centuries later the American colonists had it, too) was to anchor a barge in the middle of a swiftly flowing stream, place an axle across it, and on either end key two paddle wheels to be turned by the stream, thus driving a ponderous millstone aboard the boat. Later it occurred to some hydraulic genius that this permitted a great waste of power, since most of the river slid slyly by and did no work by impinging upon the paddle blades. The obvious remedy was to compress the river so that the power previously spread over the entire width of the stream would be concentrated against the wheel blades. Thereafter the barges were moored in narrower and narrower streams until finally the water wheel ceased to be a marine growth and became a rudimentary but businesslike land structure. The price of milled flour began to decline.

Later some ingenious Roman had another idea. The early wheels attached to barges would now be described as "undershot" because water struck their blades at the lowest point of their swing. Why not, thought an early Roman engineer, make weight, not velocity, do the work? Thereupon he devised a type of wheel which came later to be described as "overshot" because the water struck the highest, not the lowest blades. He attached to the wheel earthen jars which would fill with water at the top and spill at the bottom and thus by their weight alone carry the wheel around. Wooden troughs eventually replaced the earthen jars, but the simple principle of the overshot wheel remained the same for a considerable number of centuries. The undershot wheel likewise, after its initial development, lay quiescent, for the Middle Ages were approaching, and the Middle Ages were profuse in abbots but meager in engineers. By the time Europe emerged into the seventeenth and eighteenth centuries people knew very little more about practical hydraulics than had been known in the days of Justinian.

But the population of the world was increasing now, even though checked by wars, pestilence, and famine. No one seemed to know what might be done about the first two of these scourges, but one of the principal bulwarks against the last was the miller. As the pressure upon him to grind more grain grew greater the happy miller gladly responded by using any energy and machinery he could to produce a greater amount of milled flour out of grain. As a consequence more and more mill wheels sprang up in England and in Europe, and even in those barbaric American colonies upon which Great Britain kept a watchful eye.

Moreover, there were new stirrings in the world. The industrial revolution was beginning (for reasons set forth elsewhere) but before men could build steam engines they needed iron, and blast furnaces must supply it. Thus it was in the year 1800 that the Messrs. Crawshay installed at the Cyfarthfa Iron Works near Merthyr Tydvil in South Wales an overshot water wheel that for many years remained the largest and most powerful agent of water power in the world. It was 50 feet in diameter, 6 feet wide, had 156 buckets, and revolved at the dizzy speed of 2½ revolutions per minute. Its function was to blow bellows to keep hot the forges which supplied England's new appetite for iron.

By now the genius of England was reaching out into the far places. The Messrs. Donkin of London built in Italy a wheel 76½ feet in diameter and 2 feet wide which, with 160 iron buckets, produced the incredible amount of 30 horsepower. In the early 1800's engineers did their best to utilize accidental advantages of power like a high waterfall, and a firm of engineers tried the feat, then unexampled in complexity, of making a dual installation. At Aberdare, South Wales, they placed one wheel above another, with the lower wheel receiving water spilled out from the upper—the two wheels connected by teeth in their rims and revolving in opposite directions. Later, at Macclesfield, the firm of Chas. Wood & Bros. built two overshot wheels, one above the other, but connected by an idler-gear so that each revolved in the same direction. The efficiency of both installations was fair, but the world of hydraulic engineering was quite unconsciously waiting for some genius to tell it the true principles upon which it was to develop further.

One such genius appeared toward the middle of the eighteenth century. He was John Smeaton, an English engineer who conducted long and complicated experiments with overshot and undershot water wheels and established some of the engineering principles upon which they worked. To the Royal Society in London in May, 1759, he presented some of his results. He decided that when the head of water (that is to say, the distance through which it falls) was the same, the amount of energy produced by wheels depended upon the quantity of the water, and that vice versa, if the quantity were the same the energy depended upon the head. Two other more complicated laws he adduced also, over which the gentlemen of the Royal Society nodded sagely as if they had understood these matters all along. Actually John Smeaton was one of the chief scientific ancestors of all competent hydraulic engineers.

During the time of Smeaton's experiments there still remained only two types of wheel, the overshot and the undershot. Not long after his great work a new device began to come into use. It was known as the breast wheel, and it was the first engineering improvement in water power that had come about since the Romans had built their penstocks. Water struck the wheel a little lower than mid-way of its height and was held against the blades by sluices. The breast wheel combined the virtues of the overshot and the undershot wheel; the impounded water struck its blades so that both gravity and velocity delivered their share of energy to its axle.

These wheels began to find a wide usage during the early 1800's. Steam was beginning by now to furnish part of the world power, so much, in fact, that it was for a time to overshadow completely the less spectacular efforts of the waters. But hydraulics were not to be denied; despite the encroachments of the steam engine, the Messrs. Strutt installed, before the nineteenth century was far begun, one of the widest breast wheels ever constructed—put up at their Belper Cotton Mills. It was 40 feet wide and 12½ feet in diameter. It was a sturdy worker in the early and terrible days of the industrial revolution in England during which hundreds of thousands of people, who had previously been entirely self-sufficient for their food and clothing and shelter, suddenly found themselves flung into famine, nakedness, and destitution by their enemy, the power machine. Workers knew very little save that they were being starved to death; but the engineers of the early nineteenth century rubbed their hands in glee, for their water wheels were some 75 per cent efficient—that is, they gave out some three quarters of the energy that the water put into them. Food for a water wheel cost nothing; wheels were not subject to illness or fatigue; they would work a twenty-four-hour day, and cost little more than the interest on the large investment necessary to build them. They would deliver some twenty horsepower which was the equivalent of the daily work of two hundred men. That was none too much. The world was beginning to measure its energy requirements and find them great;

Thomas Fenwick discovered that it took very nearly one horsepower to grind a bushel of grain in an hour, and the newly frightened world of newly established capitalism began upon its first campaign of retrenchment and reduction of wages.

Meanwhile those who sought to take energy from the air were transcending mere windmills and going on to higher and finer products of the imagination less hampered by the dismal science of political economy. In the youthful United States, which had its fair quota of breast water wheels and *laissez-faire* employers, the executives of the transportation system known as the South Carolina Railway had decided in 1829 that the velocity of the wind offered them a truly God-given means of transporting their clientele to and from Charleston. Previously, these geniuses had experimented with locomotives driven by horse treadmills in the mysterious belief that a horse would deliver more energy if he were geared to the ground not by his hooves but by an inefficient belt and cogwheel machine. Disenchanted by that, they turned to sail power, still pursued by the resolute determination to spend no actual money upon the source of their energy but rather, like the Scotsman in the automat, to take what God sent. The wind locomotive of the South Carolina Railway was destined to be a magnificently inexpensive and efficient machine—so long as the wind remained squarely on its trackage and in the direction that the trains wanted to go. The steam locomotives which the thrifty management spurned were already doing good service in England but, it must be admitted, their fuel cost money. The happy South Carolina Railway paid nothing for its prime mover, and if it had only been able to school its trainmen on reef points and weather main braces it might still survive. But you cannot cross a railway conductor with a haddock fisherman, and consequently, when first the South Carolina Railway found itself running before a fresh breeze that insisted upon quartering, it got into trouble; the train had no tiller with which to luff up, and after it had reached a speed of some twelve or thirteen miles per hour the mast on the "engine" cracked, groaned horribly, and went overboard, carrying with it as many passengers as the generous sail could scoop up. Shortly thereafter the idea was abandoned, and the bankers made those explanations which they are accustomed to make.

Then, in England, the air was called upon to assist railroading by another miracle. The Messrs. Clegg and Samuda tried, both in England and in Ireland, to drive a locomotive by a different pneumatic means. Between rails, they laid a two-inch pipe with a neatly packed piston running inside. The piston carried an arm, which rose through a slot in the top of the pipe and was covered by a flexible strip of leather. Suction was to pull the piston through the pipe, and the piston was to take the train with it. It was hardly a success. Inventors were asking the wrong things of the god Æolus. When air finally came to the aid of the railroads it was to stop trains, not to start them.

Yet the idea of land propulsion by air still persisted and Mr. George Pocock in England sought another means to success. He was the gifted inventor of the "*char volante*," or kite carriage (patented in 1826), the purpose of which was to dispense with a horse and substitute for him a mammoth kite attached to a finely wrought carriage—thus to transport people from the provinces to London. His idea was not quite new. The Dutch had tried it more than one hundred years before and failed.

Mr. Pocock reported success. He also reported, for the sweet pleasure of refuting them, a number of jibes at his plans. The jibes lay less in the domain of engineering than physiology. People seemed to grant that he could transport a conveyance from Bristol to London without running it into a ditch, but they stoutly disbelieved that he could transport them at the inhuman and unnatural speed of twenty miles per hour without suffocating them. Mr. Pocock wrote a book about his kite carriages in which he insisted that he had more than once conveyed a party of people from Bristol to London by this means. He insisted also that he had delivered them to their destinations alive; the sickening rush of a twenty-mile-per-hour speed had not strangled them—although it had, he admitted, induced in all the party a strange "stomachic indifference" to lunch which he freely explained upon the basis of the unwontedly high velocity.

Although the greatest profusion of windmills still remains in Holland, it was in England that the windmill had its highest development. There it became a semi-automatic machine through the mechanical reefing and weathering devices developed by Meikle in 1750 and 1772. Great Windmill Street in London takes its name from the locality in which there stood a great mill in Tudor times. Although four arms or sails were usual then because of greater efficiency and easier construction, there still remain such mills as the 100-foot brick tower mill at Whitby, England, which has five arms, and the mill at Lewes which has six.

The tip-to-tip diameter of large arms occasionally exceeded one hundred feet and now and again the sails and hub weighed over a ton. Since the peripheral speed of sails like this often reached three thousand feet per minute, it is easy to see that Cervantes gave Don Quixote no small task to do—charging a windmill was fully as arduous as charging Richard the Lion-Hearted.

When the windmill was at its zenith there were some ten thousand of them in Holland alone, for Holland is a flat country and therefore enjoys constant winds. It is a low country also and much of the windmills' work has been in pumping the North Sea out of the *polders,* or meadows, back to where it belonged.

The western United States is still sparsely dotted with windmills to-day. But the mill in the United States which moves in the best society stands at Newport, Rhode Island. It retains only a remnant of its former grandeur, for the base alone is left standing at Truro Park. Its origin has been ascribed to Norsemen with whom, however, it had nothing to do. Governor Benedict Arnold of Rhode Island, grandfather of the Revolutionary traitor, built it some time after 1665 when a previous wooden mill had blown down. He had been born in Warwickshire, England; near by his birthplace there was, at Chesterton, one of the finest windmills in England, designed by Inigo Jones, the great English architect, who beautified whatever he undertook and who once, when he was told to build "just a barn," replied: "I will build the handsomest barn in Europe." It was the windmill of Inigo Jones which Governor Arnold copied from memory for Newport, Rhode Island, and which supplied the colonists with one of their earliest sources of power.

The windmill is not the dying machine that one might think. As late as 1903 the Royal Agricultural Society of Great Britain held a windmill contest at the Park Royal, and seventeen windmill makers submitted twenty-two mills, striving to outdo one another in the stability of the tower and the economy of its foundation, regulation and governing devices, ease of erection and maintenance, size compared to power, and last of all, price. And the windmill of that decade was some 23 per cent efficient, which means that of the energy supplied it by the inexpensive god, Æolus, it converted almost one quarter into useful work. A new type of rotor, S-shaped in cross-section and revolving horizontally, may soon do even better.

To-day most of the work of the windmill is used in pumping water—for which operation the windmill's random mixture of slothful inactivity and maniac frenzy is no disadvantage. For producing an electric current it is ordinarily useless because a constant voltage demands a constant generator speed. But anyone planning a trip to the South Pole might be wise to take a windmill in his baggage: when that gallant explorer Captain Robert Falcon Scott made his first Antarctic voyage in 1900–1904 he so equipped himself and thus made the constant, vicious winds of Antarctica supply him with electric light.

But to-day the winds do their greatest work not as sources of power but, like electricity, as transmitters of it from one place to another. If you shepherd enough zephyrs into a tank to keep them at a pressure of five to ten times that of the atmosphere, their eagerness to escape makes them quick and economical power transmitters. Perhaps the greatest single use for the harnessed winds to-day is the air brake, for having invented which old Commodore Vanderbilt of the New York Central Railroad, only sixty years ago, called George Westinghouse a fool. Furthermore, there is now not a single piece of ambitious engineering construction which could go on without the nerve-wearing and staccato assault of pneumatic machinery upon refractory stone and red-hot rivets. In this respect the elastic winds are vastly more valuable than the incompressible waters.

The use of air for transmitting power is not particularly new. About 1870 the city of Paris (perhaps because the army of the Second Empire had never been too punctual) grew tired of the great diversity of time shown by its public clocks and inaugurated a system of disciplining them by gathering them together under one system of pneumatic control. An air-driven motor is an exceptionally simple, light, and trouble-free machine, as anyone who has pumped a player piano can testify. The pneumatic clock system of Paris grew to be an astonishing success; soon a system of compressors and elaborate ganglia of piping with heavy collars and rubber gasketed joints were driving eight thousand public and private clocks in an orderly parade of minutes. Enchanted by this success, the patrons of the system soon expanded it as a means for the general transmission of power for industrial use. In 1889 it took eight thousand horsepower to drive the compressors and by 1891 a new station erected on the Quai de la Gare contemplated twenty-four thousand horsepower. In its very prime an enemy smote it. The enemy was Electricity.

CHAPTER IV

The Atmosphere Goes to Work

Suppose for a moment that you are living in the early eighteenth century and have a friend who is dissatisfied with the life of the day. He has a vision of how a million lacks of the eighteenth century might be made up. He would propose to speed up travel from twenty miles a day to one hundred miles an hour. He would replace the guttering candle you use for light with a globe that packed the brilliance of a thousand candles into the area of a ten-cent piece. Into the volume of the Sedan chair you travel in, he would cram the combined power of five hundred horses. He would see that buildings could be built to a height twenty times greater than that which stretches now from the foundation to the rooftree of your flimsy house, and that, notwithstanding, you could rise to their tops in less than a minute at no cost to yourself, or drop as quickly without breaking a bone. He would tell you that your lightest spoken words could be picked up, magnified, and flung with a power fifty thousand times greater than the combined voice of the universe across five thousand miles of barren space to be heard as perfectly as if you had whispered in a neighbor's ear. He would suggest a way through which you might walk dry-shod under rivers.

And then your visionary friend would tell you that it was not magic, not sorcery, that he would invoke to bring about these miracles. When he told you his simple means you would see to it promptly that he was cast into jail, gyved at the ankles and shackled at the wrists as all lunatics (this is still the eighteenth century) should be. For he would propose to create the twentieth century out of the eighteenth by overheating water in a closed container.

The means had always been there. Even to savages, fires were essential. They provided warmth for the shelter and means for cooking foods. In the world of work fires saw their greatest use in the smelting of metals with which tools might be made. Fires, as a matter of fact, became almost too popular; the forests of England began to be denuded for fuel, and even before the reign of Queen Elizabeth there came to be strict laws against the burning of wood. If there were to be fires, let them be made always of coal, which, like iron, could be dug from the ground and which, unlike wood, had no other uses.

The mining of coal began to be, in consequence, extremely important in man's scheme of affairs. He soon found in this a task which threatened to escape his strength. Coal, in those days, was dug from pits as open as quarries and not only ground water but rain found its way into them. Work was difficult or impossible until it was removed. That fascinating invention, the pump, helped at this task, but pumping by hand was a back-breaking task, and as pits grew larger, ran deeper, and sometimes even ducked underground to meet more seeping water, there were times when even a hundred horses could steam and snort and gallop on the treadmill and still be too few and too weak to keep down the mounting waters.

But there seemed nothing to do about it. The eighteenth century accepted the situation passively: work which had become too hard for men was now becoming too hard for horses. All the while, the fire that crackled under every pot of broth held the answer to the problem, but the century did not see it. There was no connection in sane minds between heat and energy.

Of course, that vapor rose from heated water was plain to everyone. What of it? There already existed an amusing toy which made use of this fact and anyone could see that the serious work of the world was not going to get itself done by the help of *that*. It had been invented, one hundred and fifty years before the birth of Christ, by Hero of Alexandria. A boy could build a model of it now and amuse himself by playing with this "Whirling Æolipile" if he liked. You made a closed vessel and filled it with water. Through two holes in the top you ran pillars which supported a hollow globe so that it could turn on a diameter. Two small pipes came out of the globe and were bent at right angles. One of the pillars that supported the globe you made hollow, so that steam could rise and through it could enter the globe. When all this was done, you began to heat the water in the closed vessel. As soon as steam formed, it drove itself up the hollow pillar and in a moment the right-angled pipes were spitting hot water and steam, and the globe, from the reaction, was whirling about most importantly. But that was all there was to that.

For more than seventeen centuries the Whirling Æolipile had the world of thermodynamics all to itself. Then in 1615 there came along a jostling competitor. It was the design of Solomon de Caus, a French mathematician and engineer. De Caus was not the first man, nor even the second, in the Christian world to suggest that steam could do interesting tricks; toward the close of the Middle Ages, one Malthesius had suggested some possibilities. But the identifying mark of the Middle Ages was that men talked about things without ever doing them—and when De Caus not only published "*Les Raisons des forces mouvantes avec divers desseins de fontaines*" but actually built one *dessein* of a *fontaine*, he set the scientific world once again upon a track which had been overgrown by the neglect of centuries.

Science Museum, London

HERO'S ÆOLIPILE

A MODERN model of the first of all steam turbines

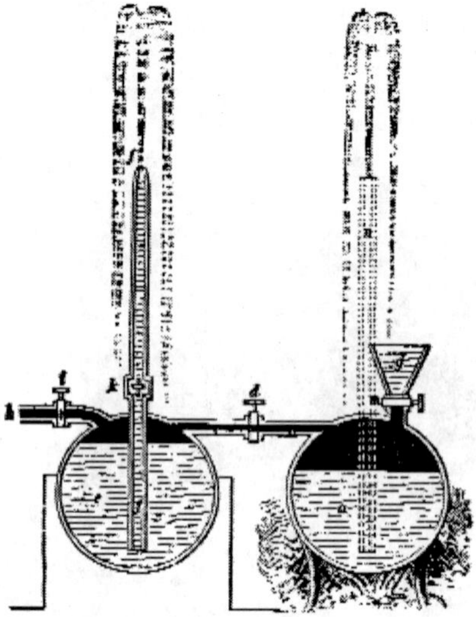

"WATER-COMMANDING" ENGINE

THE STEAM FOUNTAIN of the Marquis of Worcester, built in the seventeenth century

Smithsonian Institution

BRANCA'S DRUG STAMP

A MODERN MODEL of a seventeenth-century impulse turbine

The fountain of De Caus was an even simpler machine than the Whirling Æolipile of Hero but the world had forgotten a good deal and had to relearn it. The fountain was a globe of copper pierced by two tubes. One carried a stopcock and a funnel; the other was just a tube. De Caus poured water into the globe through the funneled tube and closed the stopcock. Then he lit a fire under the globe. His other tube projected below the surface of the water and was constricted at the top. Therefore, when the air and water in the globe were hot enough, water, in a fountain spray, would stream prettily from the top of the open tube. But that marked the end of De Caus's mechanical originality. So far as producing an engine went, he had succeeded less well than Hero.

De Caus's fountain produced its first progeny thirteen years later. Giovanni Branca, also a mathematician and a physicist, but an Italian, likewise filled a globe with water and built a fire under it. But his exit tube was in three respects different from De Caus's. It was bent at right angles, it carried a stopcock, and it did not dip below the surface of the water in the globe. Consequently, Branca could get a fairly even flow of dry steam from the mouth of the tube when his fire was hot enough. Then he showed his genius by creating, out of these materials, the first real steam engine that was ever suggested for any practical use. He blew the steam against the vanes of a wheel so that it revolved, just as the water wheels of his time revolved when a stream slid by them. And to the axle of the wheel he added a crank and to the crank a rod, and to the rod a stamp. "There, gentlemen," said Giovanni Branca, "is a machine for pounding drugs." And it was.

Now Branca's machine was a far less efficient and workable means of grinding or stamping drugs than unaided human muscles. Alchemists and apothecaries went right on as before and assured one another that no fool's work like this could ever displace the mortar and pestle. In this contemptuous way they disposed of the ancestor of all impulse turbines. But although Branca's machine was a practical failure, it accomplished one vital result: it definitely linked together in men's minds the idea of heat and the idea of energy.

And then nothing more happened for fifty years. Branca was dead and his brave drug stamp all but forgotten when Edward Somerset, Marquis of Worcester, in 1663 published his *Century of Inventions.* The idea of an engine was once again dying out of men's minds; heat as a mode of motion was no longer an unrecognized idea, but no one, for the life of him, could think what possible use such a philosophic toy might be. But then came the Marquis, diffuse, but in earnest, to tell what he had observed of the energy of overheated water.

> An admirable and most forcible way to drive up water by fire, not by drawing or sucking it upward, for that must be, as the philosopher calls it, *infra sphæram activitatis,* which is but at such a distance [sic]. But this way hath no bounder, if the vessels be strong enough; for I have taken a piece of a whole cannon, whereof the end was burst, and filled it three quarters full, stopping and screwing up the broken end, and also the touchhole, and making a constant fire under it; within twenty-four hours it burst, and made a great crack; so that having found a way to make my vessels, so that they are strengthened by the force within them, and the one to fill after the other, I have soon the water run like a constant fountain-stream forty feet high; one vessel of water, rarefied by fire, driveth up forty of cold water; and a man that tends the work is but to turn two cocks, that one vessel of water, being consumed, another begins to force and refill with cold water, and successively, the fire being tended, and kept constant, which the self-same person may likewise abundantly perform in the interim, between the necessary turnings the said cocks.

Well, the syntax of the seventeenth century is not today's; perhaps it may clear up a few puzzles for the modern mind to turn to the diagram and caption facing page 31 and see there just what the Marquis was talking about. He had built one "water commanding engine" at Raglan Castle in 1628—in wall recesses still to be seen. Another, he had built at Vauxhall, near London city, in 1647. The engines were crude enough but they worked—and had the world been ready for them they could have helped immeasurably at many tasks that bowed down the working man of the seventeenth century. As it was, the engines vanished, and only the description remained.

Then, along came Dionysius Papin, born at Blois in 1647. He was one of those restless people fated to have more ideas than are normal or convenient for any one man. In 1670 he was an assistant to the great Dutch scientist, Christiaan Huygens, at a time when Huygens was experimenting with a gunpowder engine. Later, when Louis XIV revoked the Edict of Nantes and life in France grew unpleasant for Protestants, Papin came to England and there began work with Robert Boyle on that fascinating invention, the air pump. Through Boyle he became curator of the Royal Society, and one of his duties was the preparation of a new experiment to enliven each meeting. The idea of pressures, atmospheric and otherwise, was absorbing him now, and presently his mind took an odd jump. What would happen to some bones if he cooked them under pressure in a closed container from which steam could not escape? It might make an interesting experiment for the next meeting, thought Papin; he might even prepare a philosophical supper in his "digester" and serve it up to the members afterwards.

Forthwith he set out to try. He took only one small precaution. He respected pressures; he had seen in a small way what they could do when they were not treated with sufficient deference. And he had no wish to be sprayed with steam, hot water, and jellied bones if his boiler got too hot. It was then that there occurred to him the simple little mechanism without which no steam-driven machine would dare to operate to-day: a valve in the boiler held down by a weight light enough to be lifted before the pressure of the steam became dangerously great. As simple as that. It was so that Dionysius Papin exorcised the devils from out of boilers. He was working in days when science had barely begun the battle with superstition; had it not been for him the inevitable boiler explosions that would have resulted from every attempt to operate a steam engine might have well been taken as divine warning to give up this impious inquiry into the ways of nature.

Papin's digester made the Royal Society do a great deal of thinking about the interesting properties of steam and air. Charles II, who only recently had attended a meeting of the society, chortled over the earnestness of fussy gentlemen who were spending time "only in weighing of air and doing nothing else since they sat." But no one scoffed at Papin. To the fascinated scientists of London he made the first practical demonstration of how steam could be made to produce a vacuum and how this vacuum would then be the means of doing work. On a table he set up a simple cylindrical jar that was fitted smoothly with a piston. Into the cylinder he sprayed steam until the piston reached its top. Then around the outside of the cylinder he sprayed cold water. The steam condensed and the piston was drawn sharply downward. More than that, through a cord and several pulleys the power of the machine raised a weight off the table. Here was something to marvel at. Here was a machine performing actual, useful, energetic work. Of course, thought the scientists, this is just an amusing and ingenious toy but in spite of that it is interesting. You saw the way that weight was lifted from the table?

If Papin had gone on to pursue the enormous advantage over nature which he acquired that evening he might have saved the world almost half a century of waiting. But characteristically he gave up his studies of steam in favor of something else, and left England for Italy and Germany, becoming, eventually, professor of mathematics at the University of Marburg. When, in 1705, he returned to his studies of steam, his friend, the great mathematician Leibnitz, showed him drawings of another engine and thus set him off again on a start so wrong that he was never able to retrieve his steps. Just as he was on the verge of making his name immortal he imitated instead of originating, and the waters of oblivion closed over his gifted and erratic head.

Now, for the first time the Big Business Men of the seventeenth century became interested in these suggestions of machines that might be made to do muscular work. Among the biggest of these were the mine owners. Some of them sweated as many as five hundred horses on the treadmills that manned the pit pumps, only to find that all five hundred, no matter how brutally lashed, were not equal to the work of keeping the pits unwatered. Could Edward Somerset's forty-foot fountain or Papin's moving piston be brought to bear on the problem? Along came Thomas Savery and said they could.

Captain Thomas Savery had an idea which refused to let him alone. If condensing steam produced a vacuum, as Papin had shown it did, why could not this vacuum be used to raise such water from the lower level of the pit to the higher level of the vessel and make it ready for another deep gulp of water? And better still, why not have two vessels connected so that one might be heated while the other was cooled? In other words, why not devise a pair of iron lungs alternately to inhale and exhale water?

In 1698, after long pondering, Thomas Savery took his plans to his majesty King William III. To his brusque and puzzled sovereign Thomas Savery did his best to explain his new philosophical instrument, in hope of a patent. His majesty looked, pondered, pretended to understand, and finally approved with an acceptably regal air. Then having sanctioned the idea he returned to really important matters, like the reassessment of the land tax. If mad subjects felt so strongly on the subject of boiling water it was probably just as well to humor them. The Glorious Revolution was only ten years gone by.

By 1702 Thomas Savery had published drawings and a description of his engine in a book called *The Miner's Friend* or *A Description of an Engine to Raise Water by Fire Described, and the Manner of Fixing it in Mines, With an Account of the Several Uses it is Applicable to, and an Answer to the Objections Against It.* It brought him his first opportunity to put his engine to useful work. To our modern eyes this mechanism, the dearest fruit of years of labor to Thomas Savery, seems scarcely an engine at all. It had no rods or cranks. It had not even as much as a piston. But a mine owner was interested enough to have one installed.

Savery set his engine to work at Kingston and straightway it seemed a success. It cost £50, gave a naïve suggestion of being continuous in operation, and, under favorable circumstances, would pump a few gallons of water, taking four lungfuls per minute. The mine owner was heartily pleased with Thomas Savery and his "Fire Engine," until the engine with a sudden viciousness illy in accord with its previous disposition, blew up. Neither Savery nor the mine owner had any conception of the energy which it was possible to store up in heated water, and when the boiler, being called upon to stand more than it could, pulverized itself and cast its fragments a hundred yards away, the mine owner cursed Thomas Savery and all his works, and went back to his horses. With all their disadvantages, horses didn't blow up. Savery suffered a repulse and retired for a time to learn sadly about Papin and his safety valves.

Some short time before, the philosophic problem of "raising water by the force of fire" had engaged the notice of a Devonshire blacksmith. He was Thomas Newcomen, who lived in the ancient town of Dartmouth. While his forge glowed and his anvil rang his mind was busy with the idea for a better engine. He was no scientist, but he had a clear mind and skillful hands and presently both were to serve the world with a breath-taking invention. It was called the "Atmospheric Engine."

It was the first engine which, to our modern eyes, looks like an engine. And it was the first which acted like one. Newcomen and his assistant, John Cawley, a glazier, produced it nine years after Savery's melancholy boiler explosion and after a year of secret experimenting of their own. Charles II may have been disgusted by gentlemen who weighed the air "and did nothing else since they sat," but now Papin's lesson was bearing fruit, and the air was being made by Newcomen to press down six or eight times per minute upon a piston with a pressure of 14.7 pounds per square inch and thus drive a pump. And a fascinated Mr. Back of Wolverhampton watched his mine thus emptied of water. The owner of a colliery at Griff, who had earlier been quite contemptuous of Newcomen's offer to save him £900 a year (it cost him that to feed five hundred horses) by installing his Atmospheric Engine to replace the treadmill, felt, perhaps, a moment of dismay. But nothing uncomfortable, for the Newcomen engine used an enormous quantity of fuel and was subject to frequent and puzzling spells of weakness. Even more serious trouble with the Newcomen engine lay irremediably in its design, for the pressure in its boiler had nothing whatever to do with the work that could be got out of the cylinder. Newcomen's invention was called the Atmospheric Engine because quite literally the atmosphere did the work. The expansion of the steam was not utilized. Steam did nothing save raise the dead weight of a piston—connected by a beam to a pump. It was the cylinder's vacuum, created by the condensing of the steam, which gave the atmosphere a chance to do the real work of forcing the piston down and pulling the pump up.

Moreover, the engine needed diligent attendance. It was not automatic. One vital part of the machine, not shown in the illustration, was a boy whose task was to open and close valves at proper moments. First he opened a valve to admit steam from the boiler to the cylinder. When the steam pushed the piston to the top of its stroke the boy then shut off the steam and opened another valve which admitted water to a jacket around the cylinder. This cooled and condensed the steam, and the piston sucked softly down. Now it was time for more steam. In a moment it would be time for more water. The boy's job was far from a soft snap, and upon his diligence the Newcomen engine was dependent for the number of gallons of water which it could pump out of the pits in an hour. For some time thereafter the Atmospheric Engine saw no improvement.

The next two inventions came about entirely by accident. Newcomen's engine had two great defects: its method of condensing steam was extremely wasteful and a barrier to any hope for speed—and its dependence on an underpaid and none too intelligent boy made it hopelessly erratic. A casting that was defective and a boy even more lazy than the average run of seventeenth-century youth achieved, between them, more improvement in the art of thermodynamics than all the philosophers who had schemed and planned in the eighteen centuries that had intervened since Hero first set up his Whirling Æolipile.

The casting one day served the cause of science by cracking. It was the cylinder casting, and since it was surrounded and concealed by the jacket no one knew that a long, diagonal fissure ran through it. The puzzled engineers knew only that Newcomen's engine this day was operating with a speed and a verve far above the average. The reason, which dawned some time later, was that condensing water leaked at every stroke from the jacket to the *inside* of the cylinder and, in actual contact with the steam, condensed it much more swiftly. The crack was never repaired, and thereafter when Thomas Newcomen built engines he abandoned his jacket and sprayed condensing water into intimate contact with the hot steam. In that fashion the jet condenser was born.

Humphrey Potter gets his name into history for feeling tired. He had a dull job. It was no fun whatever on a warm spring afternoon to sit by the gurgling boiler of a Newcomen engine and turn valves on and off. Steam valve on. Piston up. Steam valve off and water valve on. Piston down. Water valve off and steam valve on. Piston up. And so on, relentlessly and remorselessly, hour after weary hour. It would be far more fun to cast a hook from the bank of a near-by stream or, for that matter, to do nothing more wearing than lie and watch a pleasant cloud go sailing by. Humphrey Potter was lazy. During his round of valve-turning he wondered suddenly why the walking beam, which was doing the work of horses, couldn't be imposed upon to do some of his work, too. There was a direct relationship between what he did and what the walking beam did. Could he, perhaps, run strings from the beam to the water valve so that when the beam reached one extreme of its stroke it would operate the valve? He tried it out, and it was willing.

Accordingly, an incredulous miner one day in 1713, watching Newcomen's engine operating with a conscientiousness he had never noticed before, discovered that it was running with less than its previous complete dependence upon Humphrey Potter, who by his dislike for manual labor had brought the first crude ancestor of the automatic valve gear into the world.

Humphrey Potter lost his job. All the other boys employed as valve tenders to Newcomen's engines lost their jobs, too. Five years later one Henry Beighton made out of Humphrey Potter's strings a businesslike mechanism which definitely created the automatic engine.

So matters stayed for more than a quarter of a century. It is easy to imagine the talk of a young man of mid-eighteenth-century England who felt that the world was no longer so easy to succeed in as it had been when his father was a boy. Of course the Atmospheric Engine was a great thing, he would admit. But this Newcomen had just been a lucky blacksmith who had stumbled upon an idea and had turned it into cash and fame. There was only one such idea to a century or so: Newcomen had had his and now the Atmospheric Engine was busy unwatering fifty pits over the length and breadth of England. It was as good as it could be; there was nothing more in that idea now, and what was a young man to do who felt that the world owed him a living? And what other ideas were there? Yes, indeed, the world was harder, much harder, to make a living in than in the simple days of the early century when father had been a boy and the Stuart kings, with one lamentable interlude, ruled England.

This young gentleman was quite right: already a swirl of new ideas had made complacency a little uncomfortable. And as for the Atmospheric Engine, it was obviously successful; so successful that it continued, for fifty years after it had become an obsolete and worn-out idea, to be cherished by the owners of collieries—it had succeeded the horse in their sticky affections and what was good enough for their fathers was good enough for them. Oh, yes; the Atmospheric Engine was as good as it could be—*until someone made it better.*

The year is now 1760 and someone is about to make it better. Does it surprise anyone that the steam engine is now invented and that the name of James Watt has not yet been mentioned? It should not. He did not watch his mother's tea kettle singing on the hearth; wonder at the soft rise and fall of its lid and go out, thereupon, and invent the steam engine. He was not yet born in the year that Thomas Newcomen's engine scored its first great triumph for the delighted Mr. Back of Wolverhampton. He was in no way distinguished as a boy; and at his trade of machinist and instrument maker he was considered so mediocre that the unions of the day would not have him in their membership. In 1763 he was an obscure young man who held a position as instrument maker at the University of Glasgow. One day the model of an Atmospheric Engine which had been seized with one of its many attacks of distemper was brought to him with the request that it be repaired and returned as swiftly as possible. Watt studied this unimprovable triumph of invention and decided that he could better it considerably. He repaired the engine and returned it and, his work being over, sat down to reflect on how the invention of Thomas Newcomen could be altered to the greater pleasure of mine owners and other gentlemen who might in the future be interested in the possibility of obtaining energy from heat. That day the fortunes of our modern world began.

CHAPTER V

OVERHEATED WATER

TO HIS guild young James Watt was of small account, even though he had been no older than sixteen when he was apprenticed to a "mathematical instrument maker." This title sounds very precise and dignified. It was neither. If James Watt had had a business card in those days this is quite probably how it would have looked:

> JAMES WATT, M. I. M.
> *Glasgow*
>
> LAND SURVEYING DONE
> CLOCKS MENDED   CUTLERY SHARPENED
> MUSICAL INSTRUMENTS NEATLY ATTENDED TO
> FISHING TACKLE ALSO REPAIRED

For all these things James Watt did. There is no record that he mended umbrellas, framed pictures, or loaned money without security. But he might as well have, for the trade of mathematical instrument maker was defined only by these three things: it had little to do with mathematics, it was not particularly concerned with instruments, and it did not necessarily involve making anything. It was just a fine, resounding, and highly elastic title to indicate that a man was probably clever with his hands.

James Watt was born at Greenock, Scotland, on January

19, 1736. His grandfather had been a teacher of mathematics. His father was one of those dangerously versatile men who could be either carpenter, shipwright, builder, contractor, general merchant, ship chandler, or undertaker—and who had been all of them, although not necessarily in that order: a fine man to have on call in emergencies ranging from leaking roofs to sudden death. In his capacity of carpenter, builder, and contractor he had designed and constructed the town hall and the council chambers of Greenock. By devices not unknown in modern municipal government he had become also town councilor, treasurer, and "bailie"—this last a term by which the Scotch are still pleased to describe a certain type of municipal officer a little grander than a sheriff. In his spare time this extraordinary man had also found the leisure to become the father of five children.

Of the five, James was number four. Swiftly, in an age which knew nothing of preventive medicine, he became the eldest living as one disease or another swept his two elder brothers and elder sister away. James, his mother undoubtedly explained, was not like other boys. He was a delicate child of easy tears and a disinclination to play with the rowdies of the neighborhood. He was a pallid boy of the indoors and stayed close always to the protecting shadow of his mother, who inadvertently damned him in her sweet maternal way by saying, "He recompensed me for the loss of my only daughter."

At an early age the young James began a reluctant experiment in education at Mr. M'Adams's Commercial School—commercial since its aim was to produce useful citizens, not gentlemen. In the midst of its boisterous juvenile mob the slinking James was so sadly out of joint that his fellows had nothing but contempt for him and, most degrading of all, the girls jeered at him.

Lacking even the inspiration of one affectionate pigtail he was continuously unhappy and ill. His school work was beyond him and he spent most of his time in the corner, a helpless and terror-stricken object derisively crowned by the dunce cap. As soon as possible he would rush home and make a sad attempt to woo back self-confidence and self-esteem by drawing and whittling. He read eagerly and, to the scorn of his fellows, displayed a deep learning in wild-flowers. In only one particular could they discover that he was a normal human being: he was fond of fishing. The rest of his time he spent deep in the medical books of the day, in the study of botany, or on long solitary rambles in the woods.

Meanwhile his father had fallen upon hard times and found his jackhood in all trades insufficient to support his family. He had no great love for his odd, anemic son and in 1754, when James was eighteen, shipped him off to Glasgow to take his rightful place in the world. Did this incessant whittling mean that the boy was clever with his hands? Then obviously the best career for him was that of a mathematical instrument maker—one of those odd artisans whose place it was in the eighteenth century to fashion the crude implements which natural philosophers occasionally had need of. But in Glasgow he found his opportunities insufficient.

And so in 1755 through the shimmering curtain of her tears his mother watched him set off for London—as far away as the antipodes, she thought. But James's supposedly seven-year apprenticeship was soon cut short. He caught a cold. His seat in the shop was near the door, and the drafts were something more than his tottering constitution could withstand. He spent one year in catching fish for recreation and catching colds by profession and in the fall of 1756, homesick, discouraged, almost aggressively undistinguished and now harboring, even for him, an incredible host of influenza bacteria, he snuffled home to Greenock.

But, home again, he discovered that he was less dependent upon his parents than he had thought. A certain ambition began warring in him to become something more than a culture ground for germs. Presently a strange restlessness had driven him to Glasgow. In the first fair test with his fellows which the previous two years had given him he had learned to his surprise that he was considerably more quick and skillful than they. A determination formed in him to open a shop of his own. He did; and swiftly fell afoul of the unwritten law of his guild. The Corporation of Hammermen of the City of Glasgow undertook to teach him that no one who was neither the son of a burgess nor an indentured apprentice was to have any opportunity to indulge in the making of mathematical instruments so long as they had their healths.

Fortunately Watt had already come favorably to the notice of Dr. Dick, professor of natural philosophy at the University of Glasgow and when the guild cast Watt out, Dr. Dick, with a swiftness that was to have a profound effect upon natural philosophy, took him in. At the doors of the cloister of the quadrangle the Glasgow Corporation of Hammermen might knock in vain.

Watt took up lodging in the college apartments, which is to say the dormitories. By day he quartered himself in his shop—a little cell on the first floor whose three narrow windows looked out upon the northwest side of the inner quadrangle immediately under the gallery of the natural philosophy classroom. James Watt's clever hands and quick mind were now at the hire of the University of Glasgow, which happily "possessed funds and apparatus . . . to contribute largely to diffusion of useful knowledge among that class of individuals to whom it was particularly beneficial."

In the eighteenth century the subject of natural philosophy held in its affectionate embrace every department of what is now called science. And so it could as well have been that Dr. Joseph Black might, when Watt first began service under him, have been deeply immersed in any one of a dozen subjects. But happily, what was engaging his attention was the curious power of steam.

Dr. Joseph Black was an extraordinary teacher and scientist who all but discovered the principle of the balloon twenty years before the Montgolfier brothers and who came equally close to making great contributions to the steam engine. But his work in some peculiar way always fell just short of achieving its best ends. Many a time, however, when James Watt, lacking a scientific basis to explain some mystery, was bewildered beyond his capacity, it was Dr. Black who gave him the necessary helping hand.

Another friend was Dr. John Robison, a student from whose endlessly curious mind there came the inspiration which touched off the spark of Watt's genius. "My attention," said Watt, "was first directed, in 1759, to the subject of steam engines, by Dr. Robison, then a student in the University of Glasgow, and nearly of my own age. Robison at that time threw out the idea of applying the power of the steam engine to the moving of wheel carriages, and to other purposes; but the scheme was not matured, and was soon abandoned on his going abroad." Probably the idea was not matured because it was too big for the times. The steam engine, as Watt knew it in 1759, was a mechanism with the disposition, but without the strength, of a smallish mule; to apply it to locomotion would have been like applying a cough to the mains'l of a Dutch East Indiaman.

Yet Robison's idea took firm hold of Watt. The pallid, whimpering boy of Greenock had been transmuted now into the ablest mechanician of the University of Glasgow—pur-

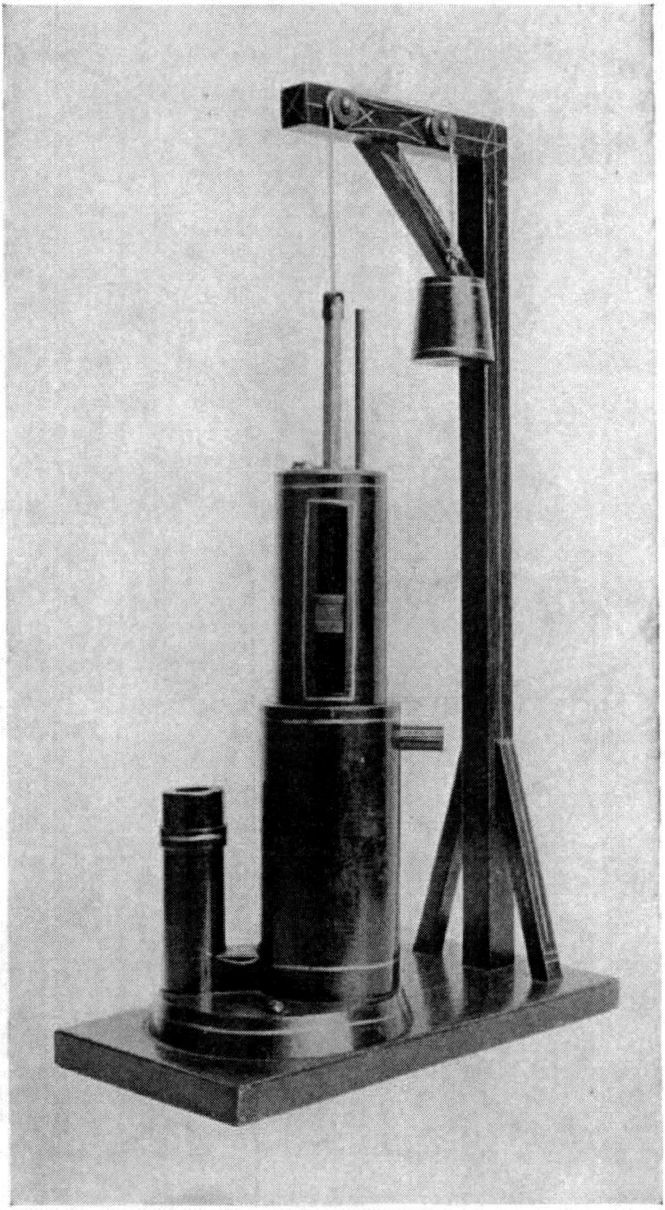

Smithsonian Institution

THE STEAM ENGINE'S ANCESTOR

A MODEL of the machine with which Dionysius Papin astounded the Royal Society

Science Museum, London

"WATER FORKER"

Existing ruins of a Newcomen atmospheric engine dating to 1754

poseful and determined to accomplish something practical in the vast but foggy realms of natural philosophy. In 1761 or 1762 (Watt did not remember which) he began his first active experiments with steam. He paid little attention to Thomas Newcomen's engine, but went straight back to the principles of Dionysius Papin and began to study the effects of steam pressure in a Papin "bone digester." Then, either to satisfy himself, or not knowing that Papin had done it before him, he built a simple piece of steam machinery similar to that with which Papin had bewildered the scientists of the Royal Society some years before. He fitted a syringe with a solid piston and a stopcock to admit or cut off steam. The piston he loaded with a weight of a few pounds, which discouraged it not at all; it rose promptly as soon as steam was admitted under it. But there for a time Watt rested, for despite the satisfaction he felt in watching his rudimentary engine do its simple work, he could not see that it was of any value unless he could form a vacuum inside the cylinder to assist the descent of the piston. If Thomas Newcomen had stupidly ignored one great possibility that lay in his Atmospheric Engine, Watt in 1762 was even more stupidly ignoring another. Newcomen had the vacuum and did not think of expanding steam as being capable of doing useful work. James Watt saw the virtues of the steam and did not know how to obtain his vacuum. If they could have pooled their ideas at that moment history might have traveled faster. But history was traveling fast enough for the eighteenth century.

In 1763 it fell to Watt to repair the model of a Newcomen engine which had been attacked by one of its frequent spells of dizziness and fainting. The engine belonged to the natural philosophy class, and had balked suddenly at its task of making evident to students the wonders of the new world of science. The immediate repairs were trifling; the real difficulty with the engine lay beyond repair. Its nine-inch boiler could not supply steam fast enough for the wasteful uses of its little two-inch brass cylinder. And so, having returned the engine as well repaired as his skillful hands could do it, Watt settled down to the line of reasoning and the mode of experiment which led him in a short, sharp drama almost straight to the truth.

Said Watt to himself, "The area of the cylinder exposes too great a surface from which heat can leak away. Accordingly the none-too-hot steam begins to condense too fast. For another thing the brass cylinder is too good a conductor of heat. For still another it is manifestly wasteful to heat the cylinder and then to cool it over and over again with every alternation of the piston strokes." Wood is a poor conductor of heat. Watt's first challenge to the problem of the ailing engine was to build a wooden cylinder six inches in diameter and twelve inches long and try to make it impervious to steam or water by impregnating it with linseed oil. But wood, thus treated, was not durable, and with this cylinder Watt could not, to his chagrin, obtain a vacuum as good as Thomas Newcomen's.

Methodical James Watt gave up the wooden cylinder idea; gave up in fact all experiments and set to work to study the properties of steam with no apparatus more complicated than a pencil, a piece of paper, and some apothecary's phials. After some months of study, the mists began to clear away. He felt ready for more experiments and constructed a boiler in which he could note the quantity of water evaporated and the quantity of steam used in a given time. He was astonished to find that a very small weight of water in the form of steam would heat a much larger weight of cold water. He found, in fact, that steam would heat six times its own weight of cold water up to the boiling point. Here was mystery on a large scale.

To only one man was this mysterious power of steam

apparent—Dr. Joseph Black. It was Dr. Black who had just developed what is now formidably known as "the theory of latent heat." He had discovered, in other words, that to heat water to a temperature of 212 degrees F. took a certain number of heat units, depending upon how cold the water was when you started. But *water* at 212 degrees was vastly different from *steam* at 212 degrees. It took a great many additional heat units to convert water into steam even though both were at the same temperature, and it was this heat energy which the steam generously returned when it was once again cooled to water.

Dr. Black went through the steps of his excellent reasoning with Watt and over Watt there burst a great light. If all this energy was necessary to produce steam then obviously it was fatal to make an engine cylinder the scene of both heating and cooling operations. The hot steam and cold water were enemies. It was best to get them as far apart as possible.

Meanwhile (the year was 1764) James Watt had acquired a wife. Having taken vows to endow her with all his worldly goods he felt a certain obligation to acquire some. Under his hands there lay a few untold billions of dollars for the future world. But to the prosaic eyes of the former Margaret Miller, his wife, they were only awkward shapes of wood and metal. Why didn't James undertake some really useful art like that of shop-keeping? And so Watt, to round out the income of a mathematical instrument maker, formed a partnership with one John Craig for the opening of a retail shop, in the salt market nearly opposite St. Andrews Street, Glasgow. Fortunately the shop was not notably successful and, feeling doubly certain that he was not cut out for a life of commerce, Watt settled down again to comfortable speculations on the steam engine.

One habit of his younger days James Watt had not abandoned. He was still fond of solitary walks. On such a walk he set out one Sunday afternoon early in the spring of the year 1765. His path took him to the Green—a grassy meadow used as a bleaching and grazing ground respectively by the girls and cows of Glasgow. To the Green on weekdays the Glasgow girls came with large kail pots in which the clothes of that strenuous day were boiled. Then the Green was dotted with young women earnestly tramping upon their blankets in the tubs and later spreading them out to dry on the tops of the grass spears. But to-day was Sunday. The Green was vacant and there was nothing to penetrate the abstracted gaze of Watt, who by the time he reached there had fallen under the grip of a great idea—how great, neither he nor anyone else in the year 1765, when America was a colony of Great Britain and France was a perfume-scented monarchy, could have had the remotest idea. Watt was thinking furiously of his steam engine, of Dr. Black and his elaborate theory of latent heat, and of the prime necessity, as he saw it, of keeping the cylinder of the engine always as hot as possible yet never permitting his condensing water to get above 100 degrees F.

What Watt was thinking he later set forth in his own words:

> I had gone to take a walk on a fine Sabbath afternoon. I had entered the Green by the gate at the foot of Charlotte-street, and had passed the old washing-house. I was thinking upon the engine at the time, and had gone as far as the herd's house, when the idea came into my mind that as steam was an elastic body it would rush into a vacuum, and if a communication were made between the cylinder and an exhausted vessel, it would rush into it, and might be there condensed without cooling the cylinder. I then saw that I must get rid of the condensed steam and injection-water if I used a jet, as in Newcomen's engine. Two ways occurred to me. First, the water might be run off by a descending pipe, if an off-let could be got at the depth of thirty-five or thirty-six feet, and any air might be extracted by a small

pump. The second was to make the pump large enough to extract both water and air. . . . I had not walked further than the Golf-house, when the whole thing was arranged in my mind.

It sounds simple, this idea; in Watt's phrasing there was no hint of the great changes it was to bring to the mode of living of every civilized man who breathed the simple air of the eighteenth century or walked in its uncomplicated sunlight. It was only this: there should be another vessel connected to the steam cylinder of the Newcomen engine; the cylinder should be always hot and the other vessel, a condenser, should be always cold. There were to be no more alternations between boiling steam and chilling water within the pardonably bewildered cylinder itself.

James Watt should have heard the voices of the oncoming centuries in his ears that night. He should have heard the crashing exhaust of the Hudson locomotive that draws the Twentieth Century Limited. He should have heard the scream of high-tension alternators flooding lands he had never heard of with more power than lay in the muscles of a million horses. He should have heard the droning hum of turbines with the power of ten thousand elephants compressed into one iron belly driving a steel ship through three thousand miles of waste water in five days. He should have seen the spires of buildings reaching up to scrape the clouds a thousand feet above the earth.

But probably he heard nothing at all beyond the faint splutter of his little spirit lamp, for it had been a hard day, this Sunday in the spring of 1765, and he was tired. In the morning he would be rested again and ready to set about the task of building this new vessel, connecting it to a new form of Mr. Thomas Newcomen's Atmospheric Engine.

The morning came, and other mornings beyond that, but the problems which had so neatly arranged themselves in

Watt's mind before he had reached the Golf-house were somehow slow to translate themselves from theory into actuality. When Dr. Robison returned to his college after a vacation, bursting unannounced into Watt's chambers in high and talkative good spirits, he found Watt staring silently at the fire, slowly turning a small tin cistern on his knee. It was this cistern that history has slowly altered into the more becomingly romantic shape of his mother's tea kettle.

Slowly Watt began to body forth his idea in prosaic terms of iron and wood and tin. He hired a cellar to the north of the beef market in King Street and here, with the help of Glasgow's blacksmiths and tinners, he built his first engine along what he felt were the proper lines. The workmanship was horrible but, ridiculous as the first engine was, it showed some sinew in its ungainly shoulders. In the excitement colds, drafts, wives, universities, and other obstacles to progress were swept aside. There must be a bigger model and Watt, who could not build anything much larger than an eight-day clock in the confines of his cellar, moved in August to a disused pottery near the Broomielaw, where he shut himself up with one assistant and set about building an engine which had a cylinder of only five or six inches diameter but the enormous stroke (that is to say, length of piston travel) of two feet. The engine was finished in two months' time. To Watt's intense gratification the piston lifted a weight of fourteen pounds.

But the engine which gave Watt such acute pleasure in 1765 was still far removed from a practical machine. It had piston, cylinder, condenser, and valve gear but it was not connected to anything in particular. The piston could be made to move up and down periodically but there was no continuous, useful work to which it could be put, save the operation of a pump—for the crank, that vital adjunct to continuous motion, had not yet been thought of in connection with this impertinent new machine. Fully as much as improved efficiency, the Newcomen engine had needed more noble uses. It was one thing to think up an improvement in the first flush of enthusiasm. It was quite another to spend laborious years in perfecting and refining and applying.

Now James Watt, no longer a care-free mathematical instrument maker but a man in the deep midst of all the sweat and struggle of the world to improve itself, found himself bedeviled by two difficulties—a lack of money and an excess of enemies.

Of the latter, one of the first was Humphrey Gainsborough, pastor of a dissenting congregation at Henley-on-Thames and brother to the painter of the Blue Boy, who claimed that he, not Watt, had originated the separate condenser. The Rev. Mr. Gainsborough had, it is true, built a model which had worked well. Cornish engineers had come to investigate it, but only after Watt had made proposals of the same sort to them. There followed a lawsuit which Watt won but which in no way prevented the Rev. Mr. Gainsborough from claiming that the condenser idea had been communicated to Watt by the "officious folley of an acquaintance who was full informed." It was the first of many troubles.

Watt now had, with his improved engine, something far beyond the dreams of Edward Somerset, or Dionysius Papin, or Thomas Newcomen. He had pushed his improvement of the Newcomen engine even further than the creation of a separate condenser. He had placed a steam jacket around the cylinder to keep its temperature constantly high. He had closed over the hitherto open top of the cylinder so that later (much later than seems reasonable) he could make the engine double-acting: that is, admit steam alternately to the top and bottom sides of the piston to produce a down stroke as powerful and positive as the upstroke, using the steam for pressure, not merely to produce a vacuum. The steam engine now ceased to be Newcomen's and became Watt's. The last vestige of the "atmospheric engine" vanished. And still Watt had no idea of where to turn for money.

Then into his life there walked one Dr. Roebuck of the famous Carron Works—an eighteenth-century armorer, possessor of men and machinery for the forging of heavy metals. Dr. Roebuck was likewise interested in the mining of coal. Among his many troubles, in this melancholy pursuit, was the presence of water in his shafts. He had either not heard very much about Thomas Newcomen's engine or else distrusted its efficiency. Roebuck, it happened, knew Dr. Black, who mentioned to him the spectacular success which young James Watt was having with models of a steam engine. Roebuck was interested. Shortly after, James Watt found himself commissioned to build his first practical engine with the duty of unwatering Dr. Roebuck's mines.

By November, 1765, Watt had sent Roebuck the drawings of a cylinder to be cast at the Carron Works and was himself supervising the dainty job of producing a piston rod at Glasgow. The latter turned out well but the cylinder was so poorly bored that it had to be discarded—the first of a long line of defective children.

"My principal hinderance in erecting engines is always the smith work," said Watt some years later. But with the cylinder rebored and construction proceeding, Roebuck's confidence grew with enormous strides even though it was taking four years to get this engine built.

But in 1767 Roebuck felt the generous impulse to pay off Watt's debts to the extent (being an innovator is an expensive business) of £1,000. In return for this kindness Roebuck was to gather to himself a two-thirds interest in Mr. Watt's improvements. Papers were passed to that effect and in 1769 the engine was completed. When tested it worked poorly.

Unfortunately the efficiency of the engine depended on preventing steam leaks between the piston and the cylinder wall. Twentieth-century mechanics do not find it overtroublesome to work to dimensions smaller than one tenthousandth of an inch, but the eighteenth-century artisans at the Carron Works were incapable of machining to a gap that a caterpillar might not have crawled through had it wished.

Watt wrapped his piston with a variety of bandages, including cork, oiled rags, and an old felt hat. Still, air leaked in and steam leaked out. Watt and Roebuck were depressed and unhappy. Helplessly, they watched the Watt engine enter upon a slow decline. The mine which it was supposed to unwater slowly flooded to a higher level than before. The rising waters bore down upon Dr. Roebuck a horrible detritus of debts, creditors, and entanglements, whereupon he went promptly and enthusiastically bankrupt. Watt was left with a joint patent and several new ideas, but no semblance of a financial backer who might make it possible for him to carry them out.

Meanwhile, as the frail strand that was Dr. Roebuck snapped, a stouter one was weaving. There was a Matthew Boulton who had a shop at Soho, two miles north of Birmingham on the Wolverhampton Road. This Matthew Boulton was a versatile manufacturer whose works produced telescopes, iron chain, glassware, clocks, enamelware and tortoise shell, and polished precious stones. For all this he needed power. Two water wheels turned his lathes and polishers well enough—except in summertime, when his streams dried up and left him demented among his curious raw materials. He had no recourse, then, save to an auxiliary power plant consisting of six to ten horses on a treadmill.

His first idea was to use a Newcomen engine to pump water up to a water wheel which would thus be turned. The enormous expense of the horses, said Boulton, "put me upon thinking of turning the mill by fire, and I made many fruitless experiments on the subject."

Boulton had, in fact, held correspondence with Benjamin Franklin (who in the eighteenth century knew everything about everything) on the subject of steam power. Boulton had met Franklin in 1758 and on February 22, 1766, had written him a letter:

> My engagements since Christmas have not permitted me to make any further progress with my fire engine; but, as the thirsty season is approaching apace, necessity will oblige me to set about it in good earnest. Query,—Which of the steam valves do you like best? Is it better to introduce the jet of cold water at the bottom of the receiver or at the top? . . . I therefore beg, if any thought occurs to your fertile genius which you think may be useful, or preserve me from error in the execution of this engine, you'll be so kind as to communicate it to me, and you'll very greatly oblige me.

To which Franklin, on the 19th of March, 1766, returned the helpful answer that he did not know.

By now, however, Boulton had met the wavering Dr. Roebuck in a business venture; through him he had learned of young Mr. Watt and his unusual engine. Once in 1767 Watt had had to make a long journey to London over the business of surveying the Forth and Clyde Canal. On his return to Glasgow he had stopped at Soho to see Matthew Boulton's great manufactory. Boulton was absent but Watt was none the less impressed by what he saw.

Meanwhile, Roebuck had failed, owing Boulton £1,200—just £200 more than Roebuck had earlier paid to Watt. To the Roebuck creditors the Watt-Roebuck engine patent was not worth a farthing. But Boulton, unlike the excellent bankers of the day, thought that there might be something in it. The patent was five years old and Watt had already had nine years of agonizing experience with his engine. "Perhaps," thought Boulton, to the extreme pleasure of the auditors, busily adjusting Dr. Roebuck's financial distress, "perhaps I am silly to let a good claim for £1,200 go by to take instead a deed to the deliriums of a crack-brained inventor. But that is exactly what I am going to do." And that is exactly what he did.

As a result a new partnership began. In 1773 the firm of Boulton & Watt hung out its shingle in Soho and announced to such portion of England as cared to listen that it was prepared to accept orders for the manufacture of pumping engines. They built one for themselves—which was called "Old Bess" and did nothing more than pump water to turn a water wheel.

Without advertising and without communications, the story of this new engine did get around. A John Wilkinson ordered that one be built to blow the bellows for his iron works at Broseley. The new firm speedily began operations. Boulton was the business man who furnished the capital, Watt the engineer who supplied the technical skill. "In the name of God fall to," wrote Boulton to Watt, "and do your best." Watt scarcely needed these urgings. He was already at work on the construction of this first really practical machine. It was 1776—and in the same year when the scratching of the fifty-six pens announced the beginning of one revolution in America, the ringing clangor of iron on iron announced the beginning of still another in England—the revolution of industry from the limitations of all the centuries thus far.

Early in 1776 the first Soho-built Boulton & Watt engine was erected in the iron works of John Wilkinson. Watt himself superintended the installation. He had almost no tools to work with. Later on one machine was to breed another with endless change and variation, but in 1776 Watt had no such luxury as side lathes, or planers, or boring tools. Everything depended upon the skillful eyes and hands of Watt and his few assistants.

Watt measured the clearance of his piston in the cylinder with a "thin shilling;" if the shilling rattled the fit was too loose. For weeks Watt hammered and pounded and filed endlessly; squinted through the rough inside of the cylinder casting; cursed a little at an apparently endless complication of getting everything arranged just so. Finally the day came when boiler and condenser were connected. Watt stoked a fire under his crazy boiler, which popped and gurgled and groaned as the steam pressure slowly rose in it to the pressure of some three pounds to every square inch. With a swiftly beating pulse he motioned his men to stand clear and opened the valve, which blew a lungful of steam against the bottom of the piston. There was a vast groan, a clanking of loose pins and bearings, and then, all of a sudden, the Watt engine moved. It was performing a useful duty in the world, and as the piston alternately closed and opened the bellows attached to it, the forges of the keen John Wilkinson glowed bright under the breath of a new force in the world.

Its fame began to spread. Throughout the placid village of Broseley there went the electrifying word that a new fire engine of hitherto unobtainable speed and power was to be seen at John Wilkinson's works. The gaping countryside crowded about the windows to see what this new wonder might be.

Even James Boswell left the side of Dr. Johnson long enough to visit Watt's and Boulton's triumphs. Boulton, who impressed Boswell as "an iron chieftain and the father of his tribe," put on his most impressive manner and roared into

Boswell's notebook: "I sell here, Sir, what all the world desires to have: POWER!"

In 1778 Watt suddenly stumbled upon the idea that had been under his nose for twenty years—and before that under the noses of everyone back to the vanished days of Hero of Alexandria. This was the idea of the "cut off" without which no steam engine would dream of operating to-day.

Probably because boilers were still flimsy affairs Watt had never believed in high-pressure steam. His idea, and that of everyone before him, had been to blow a soft breath of steam into the cylinder and to keep it blowing until the piston reached its extreme of travel. Unknowingly, Watt, like Newcomen, was discarding a great amount of power when he prodigally cast forth his steam from the cylinder into the condenser at full boiler pressure. Steam creates pressure because it is constantly trying to expand. In a flash one day in 1778 Watt realized that if he cut off his steam supply from the cylinder before the piston had traveled its full distance, he could save great quantities of steam and therefore fuel by making use of the steam's expansion as well as its amount.

He experimented by cutting off the steam supply at different points of piston travel, and soon discovered that if he cut steam off when the stroke was only one quarter completed he still got three fifths as much power from each stroke. In other words, the old way had been costing him four times as much for the production of only one and two-thirds times as much power. No steam plant could operate economically to-day had it not been for the Scotch thrift of James Watt in making this discovery so soon after he had set up his first crude engine.

So far did the fame of Boulton & Watt spread that a gentleman in Bristol, well over one hundred miles away, had notice of it and urgently raised his voice to request a duplicate. Alas, the base uses to which steam was so early put! This gentleman was a brewer who wished to be able to pump fifteen thousand gallons of ale per hour from one set of vats to another fifteen feet higher. Messrs. Boulton & Watt, blind to the approach, a century and a half later, of a Noble Experiment to render this service beyond the pale of decency or legality, conspired with the brewer's purpose. To the possible regret of some, it must be recorded that this engine likewise was a great success and enabled the Bristol brewer to make more and better ale at a higher rate of speed than before. Already the industry of the eighteenth century had begun to accelerate.

Moreover, new orders began to come in to the firm of Boulton & Watt. In London two distillers sought the aid of science and in Wales a coal mine asked to be unwatered. Both requests, quite impartially, were fulfilled by the Watt Pumping Engine.

By 1780 Boulton & Watt had made and sold forty such engines, twenty in Cornwall alone, although Cornish miners were at first skeptical, and were not convinced until they had seen the new machine "forking water" as it had never been "forked" before.

A mine known as the Wheal Virgin was flooded so rapidly that two thousand miners stood in imminent danger of losing their jobs. The Watt engine—now affectionately known as "the grasshopper" in derisiveness of its peculiar appearance when working—planted itself inextricably in the bosoms of everyone by saving the mine and with it the miners' salaries. The mine ordered five engines, and the sweating Watt wiped his brow and stood upright long enough to groan, "I fancy that I must be cut up in pieces, and a portion sent to every tribe in Israel."

Despite these successes, there seemed no money in the new

engine. Boulton was paying Watt £330 a year. To realize the ideal of his partner Boulton had, by 1785, spent some £40,000 and there had been almost no return. The riches which Boulton & Watt had so rosily foreseen remained, glittering but unobtainable, far out on the rainbow still.

Yet the firm persisted, not only in building engines but in improving them. Before Boulton & Watt began, boilers had been as simple as saucepans. It was Boulton who introduced the flue in its first simple form—a horizontal U-shaped tube, running from the firebox through the length of the boiler and doubling back again for its exit to the chimney, thereby clearly adopting it in 1780 long before the existence of the locomotive for which it was later re-invented. Here Scotch economy served an excellent purpose: the area to which heat could now be applied was much greater than when boilers were mere cans under which a wavering fire was burning. Yet they still were so flimsy that they needed not one but two safety valves. The first was an exit valve which would blow off steam beyond the maximum pressure. The second was an intake valve that would permit air to enter the boiler if the fire were removed—for Watt was worried less that a boiler would *ex*plode than that it would *im*plode, or collapse from the pressure of the atmosphere when its own internal pressure became too low.

The next improvement for which Watt yearned was some device to apply his engine to the business of rotating things —not merely moving them up and down by his piston. A reciprocating motion was ample for such simple necessities as driving a bellows or a pump. But what of a more complicated day? It was visionary, but it might come. To Watt an immediate and simple answer suggested itself—a crank and connecting rod.

Now the crank was certainly nothing new. It had been used for ages on such simple machines as spinning wheels and grindstones. There was only one trouble with the crank; no one had ever thought to patent it. In the summer of 1780 Watt built a model of an engine in which the cross-arm drove a crank and a flywheel.

In permitting his hands to move faster than his mind, Watt made his first mistake. The workmen of the Soho factory were curious. Watt could scarcely carry steam engines about with him in his vest pocket—bulk made it inconvenient. And thus, for all of Watt's attempts at secrecy a certain number of his goings on crept out into the factory beyond.

There came a Saturday night in which some few workmen of the factory repaired to the Waggon and Horses Tavern to indulge in the mild social relaxation which occasionally follows the drinking of a few glasses of ale. In this low-roofed, ivy-covered public house at Handsworth a few of them, on that Saturday evening in the summer of 1780, became a trifle boastful and one Dick Cartwright, a pattern maker, boasted himself into the center of conversation. He was working with Watt on a new engine, said Cartwright, working right with the boss himself. A great man, the boss was. A great, rich, fertile, inventive mind. Greatest mind of the century. What was more, he was working now on the best thing he had done yet. Yes, an engine that produced rotary motion out of a piston that moved only up and down.

Was that so, now, gibed the fellows, and how was that to be done? It went against all sense to believe that one motion could be made into another. Up and down was up and down, and rotating was rotating. Anyone could see that.

Whereat Cartwright wiped his foamy lips with the back of his hand and drew, with the flourish of a cavalier taking out a sword from the scabbard, a stubby piece of chalk from the pocket of his trousers. "This is how," he bellowed, and

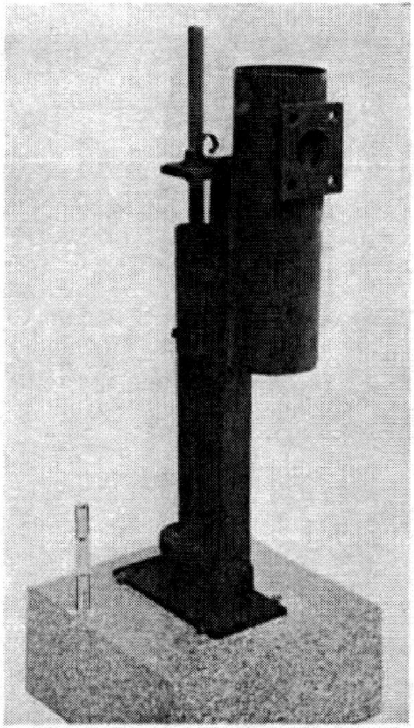

Science Museum, London

THE WORLD'S MAID-OF-ALL-WORK

THE descendants of this device make modern steam power possible. It is Watt's original model of his surface condenser

Science Museum, London

HIGH PRESSURE IN AN EARLIER DAY

A TREVITHICK-TYPE engine which James Watt bitterly fought because he feared the dangers of steam at 50 pounds pressure

Science Museum, London

MR. WATT'S WORKSHOP

STILL preserved exactly as it was found and showing that the Father of the Steam Engine felt a kinship for sculpture

Science Museum, London

"OLD BESS"

A MODEL of Boulton & Watt's first engine, ignominiously used to pump water for a water wheel

forthwith drew upon a table a slightly uncertain but quite understandable chalk picture of the linkage that Watt was using on his newest engine.

Whereat the villain of the piece, who was not evident to anyone in the Waggon and Horses Tavern, for he was dressed as a working man alike with all the others, pricked up his ears. He was there for just such breaches of discretion. Dick Cartwright and his congenial companions, quite unconscious that a vital secret had found its way to an enemy, went on with their pleasures, which led, in the ordinary routine, from cranks and connecting rods into song, from song into battle, and from battle into slumber where most convenient. Thereupon the villain had his chance at the drawing. Having pondered over it, he set out at full speed for London.

The villain of the piece was one James Pickard, a button manufacturer from Birmingham who, as only one of many eavesdroppers forever slinking about the Boulton & Watt factory in Soho, had found exactly what he wanted. And when Watt had come to the point of asking for a patent upon this new appliance, he was infuriated to discover that on August 23, 1780, a patent numbered 1263 for that year had been already granted to James Pickard, button manufacturer, for exactly the same thing.

Watt vowed stormily that he would cut off his right hand before he would pay tribute to this man Pickard whom he regarded as the thief of his idea. For a while the Watt engine continued to move up and down, but not to rotate. But then on June 21, 1781, Boulton wrote to Watt:

> The people in London, Manchester, and Birmingham are *steam-mill mad*. I don't mean to hurry you, but I think that in the course of a month or two we should determine to take out a patent for certain methods of producing rotative motion of the fire engine,—remembering that we have four months in which to describe the particulars of the invention.

Whereat Watt set to in earnest and out of his fertile mind devised five different methods for producing rotation, none of them depending upon the standard crank. All of them were obvious subterfuges. The most widely known was the "Sun and Planet Motion" with which Watt equipped a great many of his early engines. It was a clumsy evasion, but as a matter of fact the enemy of Watt was not so much the cunning Mr. Pickard as the stupid patent office. There was no more basis for the patent on a crank than for a patent on a hoe.

Meanwhile, however, the firm of Boulton & Watt had slowly begun to profit and the courageous Matthew Boulton was occasionally able to breathe a contented sigh as money began slowly to accumulate in his coffers in repayment for the £40,000 which he had poured into the firm in his stout belief that the ideas of Watt were great and would somehow prevail. When the firm first began its work it was forced to compete not only with the horses still used by conservatives but with the Newcomen engines installed by the radicals of an earlier day.

Watt had invented the term "horse power." He used it less as a scientific unit than as a sales argument. He dramatized his engine by telling mine owners and other slow-witted people what his engine could do in terms of the horses it could replace. When it came to persuading mine owners to throw out their obsolete Newcomen engines, Watt had an even harder task. Here was a machine, the first of its kind, that had been new sixty years ago; and now comes this officious young Scotsman to say that it is all out of date and that he has something better.

And so, in persuading mine owners to buy his engines, Watt found it necessary to accept payment in terms of the cost of fuel that his new engine would save. Thereupon Matthew Boulton was forced to supply sufficient cash for an engine and then wait for repayment which came in slow installments of one third of the value of the fuel that the engine saved in a certain time. To help in all these complex matters of accounting, Watt constructed an automatic counter, containing a mechanism which would show the number of engine strokes made. This, together with a record of the coal consumed, gave a figure which indicated the cost of doing work.

Each recording mechanism had two keys: one held by the proprietor, the other by the agent of Boulton & Watt. Periodically each would read the meter, and month by month the empty till of the engine makers began to fill. They were delighted. No less were the owners of the engine, two thirds of whose saving in fuel went back to line their pockets. Inch by inch the Watt steam engine edged its way into an unwilling world, not because it was a great, breath-taking, philosophic invention, but because it saved money.

But Boulton had ambitions for the Watt engine which lay beyond the unwatering of mines and the blowing of forges. Why, he wondered, could it not be applied to milling—still dependent upon winds, waters, or muscles? At this, even Watt protested; the engine seemed to him too crude and uncertain. He said as much, in the wrong quarters—and Boulton had to redouble his persuasions. He wrote in 1782:

> You must make allowance for what Mr. Watt says, he *undervalues* the merit of his own work. . . . I will take all risks in erecting an engine for a corn mill. . . . I think I can safely say our engine will grind four times the quantity of corn per bushel of coal compared with any engine hitherto erected.

He was quite right; Watt set to and designed an engine automatically controlled by a governor which he devised, and which regulated the throttle according to the centrifugal force which speed imposed upon a pair of swinging balls.

By now prosperity had come to Soho, and the two remaining thorns in life (they were considerable ones) were new patent pirates. Watt unburdened himself to Boulton in 1796. He wrote:

> The rascalls seem to have been going on as if the patent were their own. . . . We have tried every lenient means in vain; and since the fear of God has no effect upon them, we must try what the fear of the devil can do.

It was thus, apparently, that he spoke of his barristers, who in 1799 succeeded where God had failed.

Not until the first year of the dawning nineteenth century did Watt's powers and good fortunes begin to forsake him. In 1800 the Boulton & Watt patents expired. Boulton was seventy-two, Watt sixty-four. Each, however, had sons who were taking hold of the business. Then Watt's only daughter died, followed by his son, Gregory. Then Professor Robison of the University of Edinburgh, the youthful friend who had first turned Watt's thoughts in the direction of the steam engine fifty years before, and finally Matthew Boulton himself on August 17, 1809, after much suffering, at the age of eighty-one. He was buried in Handsworth Church.

Watt's chief desire now was for leisure. "My inventions are giving employment to the best part of a million people, and having added many millions to the national riches, I have a natural right to rest in my extreme age," he said. Watt has rested a long time now. He was seized by his last illness in 1819 and died on August 19th of that year, almost ten years to a day after Matthew Boulton.

But the millions of people given employment and the millions of dollars added to the national and the international riches were to be ceaselessly affected by his life. The timid boy who was the butt of school children's jokes at Mr. M'Adams's Commercial School had been transformed before he died into the Honorable James Watt, Doctor of Laws of the University of Glasgow, Fellow of the Royal Society of London, Fellow of the Royal Society of Edinburgh, Foreign Associate of the Institute of France, and Member of the Batavian Society.

Twenty years after Watt died the poet Wordsworth set down a humble tribute:

> . . . I look upon him, considering both the magnitude and the universality of his genius, as perhaps the most extraordinary man that this country ever produced: he never sought display, but was content to work in that quietness and humility, both of spirit and of outward circumstances, in which alone all that is truly great and good was ever done.

Watt's own statements were usually more simple than that. "It is a great thing," he once said to an assistant, "to find out what will not do: it leads to finding out what will do." It was in separating what would do from what would not do that James Watt spent his fruitful life, and in the course of this simple winnowing process set aside the chaff of falsehood from the kernels of truth. Few men ever more profoundly or incalculably altered the world in which they lived.

CHAPTER VI

Shining Flames

In pre-Homeric days Syrian women often wondered why their amber spindles possessed the power to attract small particles of chaff. Whirling from the distaffs, these amber spindles rubbed constantly against the spinners' garments. And it was the rubbing process, these women noted, that gave the amber its curious power. In this fashion a giant was discovered, still in its swaddling clothes.

When Greece was great, Thales of Militus (the year was 585 B. C.) first definitely set down this strange fact. Here was an interesting phenomenon, he thought, which might well bear looking into. Amber, thought Thales, must have a soul—in what other fashion could its power be explained? But Greece's studies of the amber were scanty for all of Thales, and by and by the civilization of Greece crumbled away to dust as later did Rome's. Little more had happened save that Theophrastus had grouped amber and the lodestone together because both of them had powers of attraction and Eustatius had told of a freedman of Tiberius who had been cured of the gout by the shock of a torpedo—an electric fish which both Pliny and Aristotle described. Thereafter, the world fell into the inchoate clutch of the Middle Ages and even the simplest fact that rubbed amber attracted dry paper vanished from the catalogues of human knowledge.

The Middle Ages have been explained on the basis that every mortal soul that lived between the fifth and fifteenth centuries was insane. This statement may be a trifle broad but it is, singularly, the only one which seems to account for all the multifarious but inexplicable ways in which men lived and worked and thought. Not until the twelfth century did anything even dimly relating to electricity or magnetism reappear in Europe. Then an English monk, Alexander Neckham, wrote a treatise on the mariner's compass—a device of which the Chinese in their passive and uncommunicative way had known for centuries but which was a new and startling idea in medieval England. In rapid succession, following Neckham, other similar treatises appeared.

A ferment was working now, and the world began again, intent upon exploration and discovery. So it was that William Gilbert in 1600 republished the truth that Thales had happened upon and which had lain neglected upon the scrapheap of time for more than two thousand years. Gilbert had been born in Colchester in 1540, eldest of five sons of Jerome Gilbert, the town recorder. Cambridge and Oxford both name him as a son, Cambridge with the better claim. Regardless of where he studied, he studied well. By the age of twenty-nine, he held three degrees of which the capstone was a doctorate in medicine. In 1573 he became a fellow of the College of Physicians and practised medicine in London with a success so great that Queen Elizabeth bestowed upon him her favorable notice. First he became physician-in-ordinary at her court; later, in 1600, he became chief physician to the Queen.

It was a day of vigorous ladies, and the official duties of the tall, quiet William Gilbert were not too pressing—particularly since Elizabeth, noting his extraordinary interest in the attractive powers of amber, granted him a pension to make other work financially unnecessary. In 1600 he rewarded her handsomely by publishing "*De Magnete Magneticisque Corporibus et de Magno Magnete Tellure;*

Physiologia Nova, Plurimus et Argumentis et Experimentis Demonstrata."

It was as great as it sounds. Not only was it the first valuable published book on electricity; it was likewise the first example of true inductive logic. Whatever Gilbert could prove by experiment, he included; whatever he could not, he renounced, no matter how pompous its statement or how illustrious its source. By experiment and more experiment he verified or disproved every electrical theory that had been voiced up to his time.

He had a vast amount of rubbish to clear out of the way. The Middle Ages had decided that the lodestone produced melancholy. Similarly it provided (the means was obscure) a test of female chastity. It would make a love philtre. It had "the power to reconcile husbands to their wives and to recall brides to their husbands." Toward the north there were huge lodestone mountains which would draw nails out of a ship. There were other magnets which, instead of attracting iron, would attract gold, silver, brass, lead, and "even some which would attract flesh, water, fishes." But these and all others would lose their power when rubbed with garlic, regaining it again if smeared with goat's blood—unless, of course, a diamond were near by. A diamond spoiled everything. These and other things Gilbert tested and, having proved them false, he scornfully wrote them down as "figments and falsehoods . . . put forth by raw smatterers and copyists, to be swallowed of men."

Gilbert listed all the bodies which could be made to possess an attractive power like amber. He thought them diamond, glass, opal, sulphur, resin, and mica, and adapting the word of the Greeks for amber Gilbert called them "electrics." He noted that humidity influenced attractive power, and that amber would not function well in wet weather but that on cool, dry days it might "retain the virtue" for as long as five minutes. There were some non-electrics, also, with which the virtue had nothing to do—marble, ebony, ivory, flint, emerald, pearl, coral, and all metals. Even the great Gilbert made some mistakes of his own.

Gilbert's work attracted the attention of Galileo in Italy, who wrote an early book review:

> I extremely praise, admire, and envy this author. I think him, moreover, worthy of the greatest praise for the many and true observations that he has made, to the disgrace of so many vain and fabling authors, who write not from their own knowledge, but repeat everything they hear from the foolish vulgar, without attempting to satisfy themselves of the same by experiment—perhaps that they may not diminish the size of their books.

Gilbert was the first man to consider the world a great magnet. If it were a magnet, it would have a pole. The compass needle was not, he asserted, seeking any of heaven's stars in pointing as it did, but was only aiming its delicate lance at the pole itself. So far his argument was perfect. He spoiled it by going too far and suggesting that the earth's magnetic field reached out and mingled with the moon's—that, he said, was why the earth went round.

When Gilbert died in 1603 (in the same year as his Queen) after a life more than full, Galileo himself sought to carry on his work. For five years he played with lodestones and magnets and never in all that time did he discover anything that Gilbert had not known. But Gilbert's stimulating mind produced a new result when in 1628, a quarter century after his death, Nicolaus Cabæus, a Jesuit, published *Philosophia Magnetica* at Ferrara in Italy. His religious views permitted no tolerance of vile Copernicans like Galileo and Gilbert. He set out to investigate Gilbert with all the passion that Gilbert had shown in investigating electrics. He had planned his work as an aid to religion, not to science, but it turned out the other way. He pointed out several of Gilbert's errors and in his course he happily discovered some new truths. Gilbert had said "electrics attract objects of every kind; they never repel." Cabæus showed that on occasion they did repel. He theorized (it took the world thirty years to discover his error) that a substance on the electric pushed aside the air ahead of it and then quieted down—whereupon the air would rush back, bringing the attracted chaff with it. But if he was badly wrong here, he did at least revive the use of iron filings in magnetic experiments, and thus visualized the charting of magnetic lines of force. His lines were not entirely correct, but they were a great idea for the early seventeenth century.

Thereafter learned societies began to take a new interest in the amber and the magnet. The Academy del Cimento, founded in 1657 under the patronage of Prince Leopold of Tuscany, reached two important conclusions: first, that no substances except iron and steel could intercept the pull of the lodestone; and second (Gilbert had known this but the world had mislaid it again), that electric attraction was mutual to the excited body and the body under its influence.

Things were astir in Europe now, and a few brave souls refused the mandates of authority in preference to the teachings of their own experience. In England, a few of these hotheads formed the "Invisible College" and met in cellars behind drawn curtains, lest Cromwell hang them if he caught them at the crime of thinking. In musty rooms they crowded their beards together in the candlelight and watched—*experiments.* When in 1660 Charles II came to the throne the "Invisible College" received the dignity of a charter. It became the Royal Society and Member Christopher Wren brought to light some of the views of the vanished Cabæus.

In 1663 the great mind of Otto Von Guericke produced a startling new scientific machine—a machine whereby frictional electricity could be constantly generated. Von Guericke was born at Magdeburg, Prussian Saxony, in 1602.

At Leipzig, Helmstadt, and Jena he studied law; at Leyden he learned some mechanics. Both served him well. He began a political career at the age of twenty-five as an alderman at Magdeburg, and in 1646 he became burgomaster and magistrate at Brandenburg. But over half a century of public life never dimmed his exploits in science, his recreation. His electric machine he described in a Latin volume published in Amsterdam in 1672. It was a vast improvement over rudimentary bars of glass and amber. It consisted of a ball of sulphur pierced through by an iron rod mounted on two bearings. Von Guericke would spin the axle with one hand and rub the sulphur ball with a piece of silk held in the other, and thereby obtain frictional electricity in a quantity that no one had dreamed possible before. To the wondering scientists of the day here was Aladdin's lamp come true. Said the proud inventor:

> You take the globe with you into a dark room and rub it, especially at night; light will result as when sugar is beaten. There is also a virtue of sound in this globe, for when it is carried in a hand or is held in a warm hand and thus brought to the ear, roarings and crashings are heard in it.

Thus for the first time was electricity foreseen as a bringer of light into dark places.

Another of Von Guericke's experiments clearly confirmed the dim visions of Cabæus. Electrify a sulphur globe and it would act the bully to a floating feather, repelling it when it came close enough. Von Guericke made a fascinating discovery: no matter what the position of the sulphur ball, the feather "inverts itself in the air and views the globe always with the same face."

Von Guericke played other bewildering tricks on this poor puzzled feather. If he excited his sulphur ball and held it above a table at just the right distance, the feather would first be attracted, then charged, then repelled, then discharged by the table and attracted all over again. Therein, somehow, lay the shadowy ancestor of the first doorbell.

And then this burgomaster of Magdeburg crowned his electrical discoveries. He followed up his invention of the first generator by the invention of the first transmission line. The electric flow, he found, would travel along a thread:

> This experiment clearly shows [he wrote] that the sulphur globe, having been previously excited by rubbing, can exercise likewise its virtue through a linen thread an ell or more long, and there attract something.

Then, still on the track of new discoveries, he died at eighty-four in Hamburg.

In 1675 Robert Boyle, the chemist, brought the scene of electrical experimentation back to the country in which William Gilbert had begun it. Boyle mixed chemistry and piety together in a heavy potion. His abilities soon made him a worthy successor to William Gilbert. In some respects he merely followed and confirmed his great predecessor; in others he added new facts to the world's store. He pointed out the error of Cabæus by proving that magnetism was in no way embarrassed by a vacuum—that it would no less markedly assert its force in the absence of air. He added to Gilbert's list of electrics; he was the first man to note ozone by saying "many electrical bodies may, by the very nostrils, be perceived," and the book he published in 1675 called *Experiments and Notes About the Mechanical Origine or Production of Electricity* was the first volume in history written entirely about electricity and entirely in English. Yet even Boyle, great enemy of obscure thought that he was, found it necessary now and then to be something a little less than obvious. "Why is it," a lady asked him, "that when I comb the locks of my hair they persist in flying to my cheeks?" Boyle could find nothing more helpful to answer than that it was the "effects of unheeded and, as it were, fortuitous causes." Even Thales of Militus could have done better.

By now Boyle's learning was bringing to him a great flow of scientific questioning. There came to him, for example, this letter, copied and forwarded to Boyle by John Clayton, minister in the Virginian Colony, dated "Maryland, anno 1653."

> . . . There happened about the month of November to one Mrs. Susanna Sewall, wife of Major Nicholas Sewall, of the Province aforesaid, a strange flashing of sparks in all the wearing apparel she put on, and so continued till Candlemas; and in the company of several . . . the said Susanna did send several to her wearing apparel, and when they were shaken it would fly out in sparks and make a noise much like unto bay leaves when flung into the fire; and one spark lit on Major Sewall's thumb-nail and there continued at least a minute before it went out, without any heat; all which happened in the company of
>
> WILLIAM DIGGES

Of course, no spark from Mrs. Sewall's apparel (obviously silk, and arguing her a lady of quality) ever stayed for a full minute on Major Sewall's thumbnail or anywhere else. William Digges was indulging in a human inclination which makes history hard to write—the inclination to tell a story a little taller than the happening. To this letter Clayton added one of his own:

> My Lady Baltimore, his mother-in-law, for some time before the death of her son, Cæcilius Calvert, had the like happened to her, which has made Madame Sewall much troubled as to what has happened to her. They caused Mrs. Susanna Sewall, one day, to put on her sister Digges' petticoat, which they had tried beforehand and would not sparkle; but at night when Madam Sewall put it off, it would sparkle as the rest of her garments did.

In 1684 Clayton wrote again to Boyle concerning these strange "crepitations and shining flames," saying:

> How it should transpire through the pores, and not be inflamed by the joint motion and heat of the body, and afterwards so suddenly to be actuated into sparks by the shaking or brushing of the coats, raises much my wonder.

The Rev. Mr. Clayton's wonder at the sparks of frictional electricity which cold dry silk will manifest, and which he wrongly ascribed to the person and not to the cloth, is historically important for this: it was the first observation on electricity to be made in America. It did not seem to excite Robert Boyle, who did not, this time, bother to explain it even as the result of "as it were, fortuitous causes."

By now, new observations cropped up thickly everywhere. In 1675 a puzzling paper was read before the French Académie des Sciences. The author was Jean Picard, the French astronomer whose measurements Sir Isaac Newton had used to verify his new law of gravitation. Picard reported strange doings in, of all places, the top of his barometer tube. One night when he carried it from his observatory at Paris to the Porte St. Michel he saw a peculiar glow in the Torricellian vacuum. The glow was brightest when the column of mercury descended rapidly.

Thereupon the Académie began a frenzied but unsuccessful shaking of all of its barometers. The light refused to show itself again. For a quarter of a century Picard's glow remained a mystery until, in 1700, Jean Bernouilli, professor of mathematics at Basel, discovered that if his mercury was pure enough and his tube without the suspicion of moisture, he could produce the light at will. The Académie had had no faint suspicion of any connection between the force that attracted amber and the glow in the mercury column. Bernouilli linked them together by showing that a vacuum was unnecessary; when air was present and the mercury was shaken sparks "arise simultaneously and perish at almost the same time"—certainly a forceful way of saying that they were short-endured—whereas in a vacuum "the light is like a continuous flame which lasts incessantly while the quicksilver is in agitation." Here then, said the Académie, is a new illuminant that will soon displace the candle. Ah, yes indeed, gentlemen; quite soon by contrast with the time that candles had been the only source of light. In, to be exact, one hundred and seventy-nine years and some few days.

Francis Hawksbee (spelled also Hauksbee) promptly picked up Bernouilli's clues and between 1705 and 1711 several times startled the Royal Society of which he was curator. He let air rush through mercury into a vacuum, carrying the mercury with it and dashing drops of it to every side of a glass vessel. It rewarded him, he said, by looking "like one great flaming masse." He began using larger and larger flasks and more and more mercury and when once, in a fit of trembling excitement, he shot three pounds of mercury into his receiver at once, "the light darted thick from the crown of the including glass like flashes of lightning." Here, in Hawksbee's words, was an early suggestion that the electric fire and the lightning of a storm had anything in common.

Hawksbee redoubled his experiments. Could amber be made to act as the mercury did? He built a machine to revolve an amber ball in a vacuum in contact with wool, and behold, just the same sort of light appeared. It became brighter as he increased the vacuum. But the amber charred and the wool scorched. Hawksbee then built himself a machine like Von Guericke's, but with a hollow exhausted glass globe in place of the ball of sulphur. He produced a light so bright that "large print, without much difficulty could be read by it and at the same time the room, which was large and wide, became sensibly enlightened, and the wall was visible at the remotest distance, which was at least ten feet." The Royal Society saw and became most unscientifically excited, as well it might.

Yet thereupon, curiously enough, there came a pause in progress. Sciences, like empires, have days of greatness and days of decline. Soon after Hawksbee the study of the electric fire died down largely because the placid eighteenth century felt quite certain that there was nothing more about it to discover. Had not Von Guericke invented a powerful machine and sent a charge over a thread? Had not the great Sir Isaac Newton dallied with electricity and said, "I have been much amused by ye singular phenomenon resulting from bringing of a needle into contact with a piece of amber or resin fricated on silke clothe?" Could anyone hope to add to the discoveries of Boyle, Picard, Bernouilli, and Hawksbee? Obviously, it would be more fruitful for scientists to turn their attention now to less thoroughly studied matters.

Most of them did. Only Stephen Gray, pensioner of the Charter House School and fellow of the Royal Society, decided to follow further the paths of his eminent predecessors. As a result Stephen Gray, with his famous Charter House School experiments, laid down in about the year 1720 the foundations of a true science of electricity.

Gray began by following out the studies of Von Guericke, using Sir Isaac Newton's improved friction machine with a glass cylinder instead of a sulphur ball. He stood boys (no doubt to the great distaste of their mothers) on cakes of resin to insulate them from the ground, and then charged them full of electricity by revolving his glass cylinder between their hands. Here seems to have been the first storage of the "electric fluid" outside of the cylinder itself—providing we except the crepitating Mrs. Sewall. When a boy had been sufficiently charged, he could, by touching another

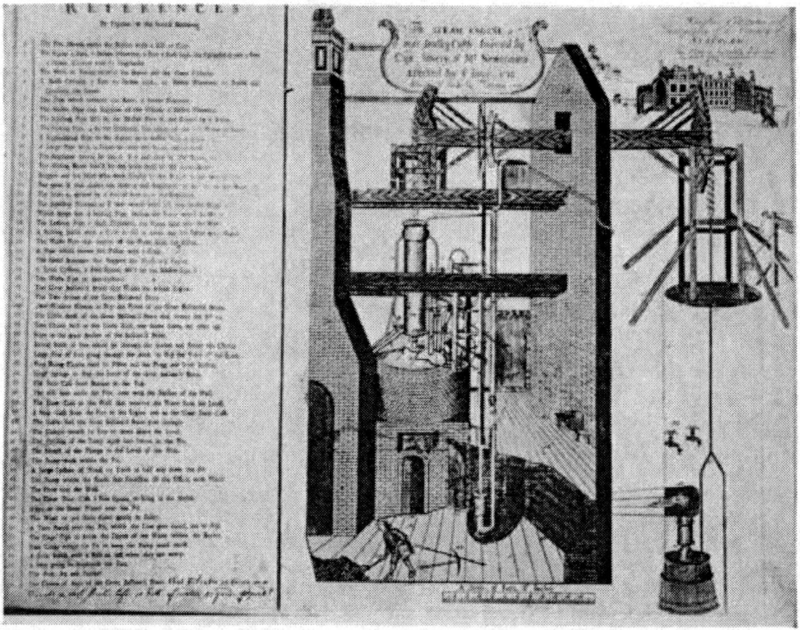

Science Museum, London

NEWCOMEN'S ENGINE

AN EARLY DRAWING of a pre-Watt steam engine erected near Dudley Castle in 1712

THE BEGINNINGS OF ELECTRO-THERAPY
Medicina electrica being wreaked upon a lady of an earlier day

boy not insulated by the resin, produce a tiny spark and feel a minute tingle. But where did the virtue go? Gray began to find out about that. One day it amazed him to discover that not only would an electrified glass tube attract feathers —so would the corks which he had placed in each end of the tubes to keep out dust. Well, then, if the virtue would travel from the glass to the cork would it travel from the cork to something else? Stephen Gray speared the cork with a four-inch wooden rod, an ivory ball on its other end. When he rubbed his glass tube the ivory ball promptly drew to itself a small bit of metal foil.

Now Gray was hot on the trail of a new discovery. He increased the length of his wooden rod. He substituted a wire for it. Then he substituted a thread for the wire. The virtue cared not a whit for all these changes; the ivory ball always proved as attractive to the metal foil.

Here was something great. Stephen Gray had rediscovered the principle of conductance; discovered that the electric flow could be given direction and made to travel in set paths over certain distances by the use of wires and threads. What would happen if he became really ambitious and sought to make the flow travel not merely for a few inches but for many feet? He reeled out a transmission line and suspended it by cords of hemp looped over a beam.

Gray's excitement changed to unhappy discouragement when he found that no matter how hard he ground away at his friction machine the little ivory ball at the end of his line showed no attraction whatever for metal leaf, chaff, feathers, or paper. To his associate Wheeler, a member of the Royal Society, Gray told his disappointment. Both pondered together and decided that suspending a hemp transmission line by more of the same hemp might be the cause of the trouble. Perhaps the virtue was leaking away through the large diameter of the hempen suspending cord. Very well, then, they would try silk because it was small and strong and "less virtue would probably pass off by it than by the thickness of the hempen line which had been previously used." Gray tied up the transmission line again, connected it to the friction machine, placed a bit of metal leaf about eight inches away from the ivory ball, seized the crank of his generator and gave it a mighty swirl. Instantly the fluttering leaf sidled over to the ivory ball and clung to it tenderly. Triumph!

Yet still Stephen Gray had not learned his lesson. Metal wire was like silk, strong but thin. So he substituted wire for the silk and churned his friction machine again, only to discover that no amount of effort could change the indifference of the metal leaf for the ivory ball. Then, finally, the truth dawned. Materials were not only electrics or non-electrics as William Gilbert had said; either they contained the virtue (insulators) or else they cast it prodigally away (conductors). This fact finally established, Gray and Wheeler in June, 1729, and August, 1730, succeeded in transmitting the virtue (they had begun with distances of less than eighty feet) as far as 765 feet through silk-suspended hemp and 866 feet through silk-suspended wire.

Gray now became a constant and enthusiastic producer of sparks. There cropped up in him the same idea earlier noted by Hawksbee and by Newton:

> And although the effects are at present but *in minimus,* it is probable, in time, there may be found out a way to collect a greater quantity of the electric fire, and consequently to increase the force of that power which by several of these experiments, if we are permitted to compare great things with small, seems to be of the same nature with that of thunder and lightning.

Stephen Gray died in 1736, one of the first electrical engineers of all time.

In the year of Gray's death the Scotch monk Andreas Gordon, professor of philosophy at Erfurt, Germany, suddenly invented the first machine to utilize electricity. Von Guericke had generated it, Stephen Gray had transmitted it, and now it was obviously up to someone to use it. The monk Gordon used it to ring a bell.

It was not much of a bell—nothing that would have called a busy eighteenth-century housewife from her loom to her door. It consisted merely of two light gongs between which hung a small metal ball suspended by an insulating silk thread. When Gordon charged the ball by touching a rubbed glass rod to it, it was promptly attracted to one gong. It struck, tinkled, lost its charge, swung back and struck the other gong. One faint *ding*, one faint *dong*.

Now what earthly use was that? Institutions of learning in Europe were grown-up places; even John Harvard's "Colledge" in the barbarous American colonies was an ancient institution that had been established for one hundred years. Where was the world ever to get, one can imagine the hard-headed business men of the eighteenth century asking themselves, if the teachers of youth, particularly an illustrious one like the professor of philosophy at Erfurt, were to waste their time in idle play like this? There was much discussion and some criticism of the monk Gordon's ding-dong bell.

Undiscouraged, Gordon went on with his silly play. He created an immense improvement in the machine of Von Guericke by using a glass cylinder instead of a globe and by using a bow (on the same principle as the American Indian's fire bow) to drive the machine's axle. Eight inches long and four inches in diameter his cylinder was, and by means of his driving bow he was able to achieve 680 revolutions for it per minute—faster, perhaps, than any piece of man's machinery had run before. Then, having enormously improved his generating facilities, the monk Gordon went on to create another electrical utility by inventing a motor.

Like his bell, the motor was not much of a machine and

had no progeny immediate or remote. But it did work. It worked on the reaction principle of Hero's "Whirling Æolipile"—that which had spat steam and water. The motor revolved from the spitting of sparks. It was nothing more than a light metal star, delicately pivoted at its center and with its points bent slightly to one side. When the monk Gordon directed at the star points a wire connected to his friction machine and sawed hard at his bow, a spark would jump from wire to star point. The kick of the sparks, as one followed another, drove his tiny star around. More foolishness!

But even the bell and the motor did not exhaust Gordon's fertile mind. The first electrical bell ringer, the first electrical motorman, became also the first electrical executioner. The monk Gordon caught chaffinches and sent them into eternity with a darting spark from his electrical machine. He did this as a philosophical diversion only and with no thought of all the benign possibilities that lay in the future. Chaffinches are sizable birds; unless he was indulging in mere secular exaggeration, the monk Gordon's friction machine must have been producing sparks of an altogether new magnitude. And his multifarious adventures set many another experimenter off into new demonstrations of what the mysterious force could do. In Europe men began to ignite alcohol and fire guns by electric sparks. Probably there was never a time in the world when the activities of natural philosophers seemed to the common man less sensible; never a time when tradesfolk or nobility, or perhaps even the poets, thought men of science more hopelessly childish. Whirling stars, tinkling gongs, and burning alcohol indeed! To say nothing of dead chaffinches! Well, it was a lucky thing that there were still good stout citizens who would travel the seas in sail, mine coal for the grate, dig iron to be smelted, fight battles with blunderbusses—and withal keep their wives subjugated to weaving and spinning and churning and baking. For what would ever become of the world if there were only silly dreamers and playboys in it like Stephen Gray and the monk Gordon and even those learned fellows, Gilbert and Galileo? Things would come to a sorry pass indeed with no one to do the work.

CHAPTER VII

The Virtue of the Phials

So far, the triumphs in electricity lay with the English. The orderly mind of Charles François d'Asternay Dufay (he was supposed to spend his time as superintendent of gardens for Louis XIV) soon produced a commotion in France. He decided from what he had observed that there were two kinds of electrical charges. If, as Cabæus had shown, electrics sometimes attracted and sometimes repelled, it could only be because they had been charged in different ways.

What, he wondered, would a piece of gold leaf do under the action of *two* electrified bodies? He investigated and found that electrified glass repelled the leaf. Glass, then, was in a class by itself. It repelled, whereas repeated experiment showed that electrified gum copal, like amber and Spanish wax, attracted.

> Electricity [said Dufay in 1737] is a quality universally expanded in all the matter we know, and which influences the mechanism of the universe far more than we think.

He experimented with every possible combination of electrics and came to a conclusion. If, said Dufay, you rub glass with silk, you produce *vitreous* electricity. And if you rub resin or rubber or amber or sulphur with fur, you produce *resinous* electricity. Crude though this was, it was a help. It classified things and arranged them. Before long Dufay's vitreous electricity came to be known as *positive,* and his resinous to be known as *negative.* When one considers that the latest scientific talk on the still baffling question says that positive electricity is Nothing and that negative electricity (which should have been called positive but got so mixed up years ago that nobody has ever been able to straighten it out since) is a "flow of electrons," one will see that Dufay was telling the eighteenth century great and penetrating things. He went on to enlarge other theoretical ideas, but his greatest achievement was perhaps the sending of a charge over a wet string 1,256 feet long; a distance considerably greater than Stephen Gray had been able to accomplish.

Dufay had ambition to collect a large quantity of electric fire. The eighteenth-century mind was not good at abstraction, and since electricity consisted, obviously, of the flow of something from one place to another, then electricity, thought everyone, must be a fluid as much as water—a substance by some curious chance invisible, and lacking weight.

This idea was mistaken but fortunate. It set scientists everywhere to stalking the electrical substance, with ambitions like Dufay's to collect as much of it as they could in one place. This ambition led Pieter Van Musschenbroek to an important and uncomfortable discovery. In Leyden, Holland, the good Pieter, who was professor of natural philosophy successively at Duisburg, Utrecht, and Leyden, decided to bottle some electrical fluid by charging the water in a jar with a friction machine. He connected the two together and in the semidarkness of his laboratory he ground and ground away at his electrical machine. Finally, he decided, he had enough.

He certainly had. He had far more than he wanted. He had built up many thousands of volts of static electricity in his insulated jar and when he rashly sought to touch it, the darkness of his room was lighted by a livid blue flash and Pieter Van Musschenbroek next came to consciousness sitting dazed in a corner with his spine badly jarred and his stout

Dutch dignity fractured altogether. He was not a man to take an insult like this lightly. He cursed and further soothed himself by smashing things. But wisely he refrained from touching the jar, the innocent little jar, still seated so comfortably on his table.

Yet later, in January, 1746, he did repeat the experiment, this time with Professor Allamand and one Cunæus, a rich burgher of Leyden. Now, however, he ground less ambitiously at his friction machine. Once again the results were startling although the jar was a trifle more restrained. Pieter wrote to his scientific friend René Reaumur:

> I wish to inform you of a new and terrible experiment which I advise you on no account personally to attempt. I am engaged in a research to determine the strength of electricity. With this object, I had suspended by two blue silk threads a gun barrel, which received electricity by communication from a glass globe which was turned rapidly on its axis by one operator, while another pressed his hands against it. From the opposite end of the gun barrel, hung a brass wire, the end of which entered a glass jar, which was partly full of water. This jar I held in my right hand, while with my left I attempted to draw sparks from the gun barrel. Suddenly I received in my right hand a shock of such violence that my whole body was shaken as by a lightning stroke. The vessel, although of glass, was not broken, nor was the hand displaced by the commotion: but the arm and body were affected in a manner more terrible than I can express. In a word, I believed that I was done for. For the whole Kingdom of France I would not take a second shock.

What he had done for science was to make available the first crude form of condenser. What he had discovered for himself was that the condenser, being otherwise insulated, had, through his courtesy, discharged itself into the ground. As a matter of fact, he suffered needlessly, for although the experiments remained for some time unknown, Bishop Von Kleist, dean of the Cathedral of Camin, in Pomerania, had completed a similar work on October 11, 1745, and through a friend reported it to the Academy of Lignitz. He had written this:

> When a nail or a piece of thick brass wire is put into a small apothecary's phial and electrified, remarkable effects follow: but the phial must be very dry or warm. I commonly rub it over before hand with a finger, on which I put some powdered chalk. If a little mercury or a few drops of spirit of wine be put into it, the experiment succeeds the better. As soon as this phial and nail are removed from the electrifying glass, or the prime conductor to which it has been exposed is taken away, it throws out a pencil of flame so long that, with this burning machine in my hand, I have taken sixty steps in walking about my room; when it is electrified strongly, I can take it into another room and there fire spirits of wine with it. If, while it is electrifying, I put my finger, or a piece of gold which I hold in my hand, to the nail, I receive a shock which stuns my arms and shoulders.
>
> A thin tube, or a man placed upon electrics is electrified much stronger by this means than in the common way. When I present this phial and nail to a tin tube which I have, fifteen feet long, nothing but experience can make a person believe how strongly it is electrified. Two thin glasses have been broken by the shock of it.

Von Kleist's discovery probably preceded Van Musschenbroek's, but no one then cared much about finicky questions, and the phial came to be called by the name of the town in which it had first been heard of—the Leyden phial, later to become the Leyden jar. And a short time afterwards Sir William Watson, scientist and apothecary of London, and Dr. John Bevis gave it its present form by dispensing with the water, the gun barrel, the mercury, and the spirit of wine and making it simply an empty bottle covered inside and out to about half its height with tinfoil.

The Leyden phial and its accumulated virtue swiftly became the most important instrument in the hands of electrical experimenters. Tiberius Cavallo, one of the eighteenth century's great historians of science, bowed low before it in 1795. This is how he paid it his respects in the course of a ponderous, three-volume treatise on electricity:

> Its [electricity's] attraction could in part be imitated by a lodestone; its light by phosphorus; and in short nothing contributed to make electricity the subject of public attention, and excite a general curiosity until the discovery of the vast accumulation of its powers, in what is commonly called the Leyden Phial, which was accidentally discovered in the year 1745. Then, and not till then, did the study of electricity become general, surprised every beholder, and invited to the houses of electricians a greater number of spectators than were before assembled together to observe any philosophical experiments whatever.
>
> . . . Discoveries crowded upon discoveries; improvements upon improvements; and the science ever since that time went on with so rapid a course, and is now spreading so amazingly fast, that it seems as if the subject would soon be exhausted, and electricians arrive at the end of their researches; but, however, the ne plus ultra is, in all probability, as yet at a great distance, and the young electrician has a vast field before him, highly deserving his attention, and promising further discoveries, perhaps, equally, or more important than those already made.

Through the commas it is possible to discover that Cavallo thought quite well of electricity and was visionary enough to think that the year 1795 might not, after all, bring the end of its advances. But what were some of the crowding improvements that followed the discomfiture of Pieter Van Musschenbroek?

For one thing, the Abbé Jean Antoine Nollet in France assembled a company of the King's men, joined their hands, and then in a very secular fashion knocked 180 of them down with one heavy jolt from a highly charged phial. Later, at a Carthusian monastery he sent a shock through a line of monks stretched hand in hand for over a mile. In England, the less frivolous Royal Society utilized one of its committees to send "an electrical commotion" from a "charged phial" over a transmission line. How much virtue could they get from it, they wanted to know, what distance could they transmit it, and how long would it take to get there?

To answer these weighty questions, the Royal Society designated Dr. (later Sir) William Watson, assisted by Martin Folkes, president, Lord Charles Cavendish, Dr. Bevis, Mr. Graham, Dr. Birch, Peter Daval, and the Messrs. Trembley, Ellicott, Robins, and Short. On July 14 and 18, 1747, these gentlemen began sober experiments on the banks of the Thames at Lambeth, and transmitted the virtue to the opposite bank across Westminster Bridge. On July 24th at New River, Stoke Newington, they went one better than Stephen Gray. They sent a charge of virtue through 800 feet of water and 2,000 feet of land. This accomplishment was triumph enough, but on the same day the Royal Society tasted one even sweeter. It sent the virtue through 2,800 feet of land and 8,000 feet of water—to a total distance of a little over two miles. The pleased Dr. Watson could not discover the transmission time which, by all his instruments, was nothing at all. Since we now seem to know that electricity travels 186,000 miles in a second, it is not to be wondered that Dr. Watson, with his hour glass and whatever other "philosophical instruments" he may have used, was unable to measure the one hundred-thousandth of a second involved.

But he was happy and delighted just the same. A year later (August 5, 1748) he made further experiments, this time at Shooter's Hill, and set up a new, startling long-distance transmission record—12,276 feet. Once again, the time defied his sun dial.

Thereupon Dr. Watson set out to apply electricity to every human use. He passed it through a vacuum. He applied it to medicine. A seven-year-old girl in the Foundling Hospital was seized with a disorder, possibly epileptic, of which the alarming and discommoding symptom was "universal rigidity of muscles." She seemed all but dead, and medicines were ineffectually administered for a month. Mallow leaves, beet root, aloes, nutmeg, dissolved pearls, rue, angelica—the eighteenth-century pharmacopœia was helpless. Then came Dr. Watson with the friction machine and the Leyden phial and set up the science of Electro-therapeutics. He electrified her intermittently, and at the end of two months, rather than stand it any longer, she got well, and could "exercise every muscle in her body without pain." The merciful means of cure may be seen in the illustration facing page 79.

Thereupon, electricity became the great cure-all. Dentists, then as now keen to discover any new means of torture, used the new means to transmit virtue into aching teeth. Two wires, adjustable to fit all sizes and shapes, "embraced firmly" the tooth and were then connected to a charged Leyden phial. Happy dentists reported that a single shock would often cure an ache instantly, and that two or three were invariably successful. The suspicion was strong that any normal patient would gladly endure a toothache in stoical silence all the rest of his life rather than submit to more shocks than three. For similar mercies William Watson was knighted in 1786, one year before his death.

There now enters upon the scene of electricity the great Benjamin Franklin, illustrious Amateur of Everything, reputedly youngest son of a youngest son for five generations, who found time in his busy life to be a printer, a man of letters, a diplomat, a statesman, and a scientist.

To Boston in 1746 came one Dr. Spence, lately arrived from Scotland. It was the year of the Leyden phial, and its stored virtue was the great discovery of science. Dr. Spence and Franklin met, and the doctor showed Franklin his crude apparatus and performed a few experiments. All this was new to Franklin and he turned his fascinated attention to it. He stood men on cakes of wax as Stephen Gray had once done and did his best to calculate the amount of electrical discharge from them under varying circumstances. Through a variety of enthusiastic experiments, Franklin evolved a theory: uncharged bodies contained a fluid and this fluid was electricity. When it was present in its normal amounts it produced no effect. But an excess of the fluid produced a positive charge and a deficiency produced a negative. And the particles of fluid, he thought further, repelled each other but attracted ordinary matter. He electrified a jet of water, watched it break into a spray as a result, and took it to be a confirmation of his theory. These theories were later to bring Franklin into conflict with the great French scientist, Coulomb, who was responsible for the first precise electrical measurements and who was to succeed in determining the force between two small charges by means of a torsion balance. Electricity is two fluids, said Coulomb. Electricity is one fluid, said Franklin. Each, by the beliefs of the present, was wrong. The quarrel grows less simple every day.

Yet Franklin was far from confining himself to theory. He was an unquenchable experimenter. He showed, for one thing, that a charged Leyden jar *contained* nothing new—the distribution of fluids was the only thing that made it charged; when he connected the inside of the jar to the outside, the prompt flash discharged it as completely as contact with the ground. The inner coating of the jar was positive; the outer one was negative. But just where did the mysterious and elusive fluid store itself? The scientists of Europe believed that it was held in the water, the iron filings, or the metal foil of the phials. Franklin showed, one after another, that these were not its lairs. He finally tracked the electricity down to its hiding place—the place so easy and obvious that no one had ever thought to look for it there. The electricity smiled at him from the surface of the glass!

There was a simple corollary to this: if the electricity was on the glass then the capacity of the jar depended on its area and two jars would be twice as powerful as one. Practice sided with theory when Franklin made a battery of six jars and connected them to a turkey. One can see that picture: the forty-year-old printer letting Poor Richard's Almanac go hang while he studied the virtue in Leyden phials; the turkey, righteous and distant, as proud as a dowager at the opera, but with a cold film of distrust in its eye, remotely watching while Franklin spun his friction machine to pour into the Leyden phials his mysterious and invisible fluid of death. It was no way, perhaps, to treat the great American bird, but it was in the cause of science, and a scientist on the trail of a theory respects neither God, nor man, nor turkeys.

In 1749 the same idea occurred to Franklin that had earlier occurred to Hawksbee and Newton: that there was something in common between the virtue of the phials and the lightning of the sky. The same Abbé Nollet who shocked his fellow monks had written three years earlier:

> If anyone should undertake to prove . . . that thunder is in the hand of nature what electricity is in ours—that those wonders which we dispose at our pleasure are only imitations on a small scale of those grand effects which terrify us . . . I confess that this idea, well supported, would please me very much.

But where all others had previously suggested the similarity and done nothing to prove it, Franklin set about to see definitely whether it was so. The electric spark, he said, travels a zigzag course. So does lightning. The electric spark jumps quickest to pointed objects. Lightning strikes trees and church spires. Electricity chooses metal conductors. So does lightning. Electricity ignites combustibles. So does lightning. Electricity reverses the poles of a magnet. So does lightning. They must be the same thing.

But were they? Franklin's first idea was to erect a sentry

box on a tall hill and from it project a pointed iron rod stuck in resin. To that, he would present his key and his knuckle and feel the current drawn off from the electrified clouds. Later, he thought of the advantages of a church spire. But he could find none suitable, and for three years his experiment got no further than the back of his mind.

In 1752, the history-making idea came to him—the idea of flying a kite into the clouds. In June he made his experiment. He was inviting death; perhaps ignorance made him brave. With his son, he raised his kite from the Philadelphia Common. There he stood, with clouds rolling over the city and a drenching rain on his silk kite, which dipped and fluttered in the dark sky. At the end of the kite string there dangled the famous key. For insulation his hand was wrapped in a silk handkerchief. So long as the string was dry, nothing happened. But when the pelting rains sufficiently wet it, a spark promptly jumped to his diplomatic knuckle and, jumping, established one more fact in science's widening horizon.

If Franklin's luck had not been good, someone else would have been the first American ambassador to France. There is no particular reason why he was not killed. Yet having established his fact, he thereupon devised the lightning rod, which shortly pierced its lance into the sky from almost every rooftree in Europe and America. Lightning was somewhat more of a fright to the eighteenth century than it is to the twentieth, although there are still nervous ladies who, at the first hint of a summer cloud, lie down on a feather bed supported in glass castor cups and defend themselves against a possible Act of God by a fence of rubber boots. And this was how Benjamin Franklin suggested that the scourge of lightning might be abated:

> It is still safer to bring two or three mattresses or beds into the middle of the room, and folding them up double, put the

chair upon them; for they, not being so good conductors as the walls, the lightning will not choose an interrupted course through the air of the room and the bedding. . . . But where it can be had, a hammock or swinging bed suspended by silk cords, equally distant from the walls on every side, and from the ceiling and floor above and below, affords the safest situation a person can have in any room whatever, and what indeed may be deemed quite free from danger of any stroke by lightning.

This was esteemed a trifle impious; there was an earthquake in 1755 and the pastor of the Old South Church in Boston insisted, in a loud sermon, that it was God's wrath—directed at Franklin for attempting to ward off His vengeance with the lightning rod.

But nothing dimmed Franklin's zeal for electrical experiment. He was the greatest showman of his day. If electricity had been new this morning, no ingenious press agent could have thought up more ingenious stunts than Franklin suggested then:

The summer coming on, a bad season for electrical experiments, it is proposed to put an end to them for this season, somewhat humorously, in a party of pleasure, on the banks of the Schuylkill. Spirits at the same time are to be fired by a spark sent from side to side through the river without any other conductor than the water—an experiment which we sometime since performed to the amazement of many. A turkey is to be killed for our dinner by the electric shock, and roasted by the electric jack, before a fire kindled by the electrified bottle, when the healths of all the famous electricians in England, Holland, France, and Germany are to be drunk in electrified bumpers, under the discharge of guns from the electrical battery.

A good electrified time was had by all. The standards of that day being somewhat different from the present, the effect of the electrified bumpers was to send a great number of the company, including good Benjamin Franklin himself, home, rich in that virtue which no ordinary phial possesses.

Photo Giraudon

"VOITURE A VAPEUR"

A MODEL of Cugnot's steam cart—second steam-propelled vehicle in history

Smithsonian Institution

TREVITHICK'S FIRST ATTEMPT

A MODEL of the simple steam-driven velocipede which was Richard Trevithick's first contribution to locomotive design

Baltimore & Ohio Railroad Archives

FIRST ON THE RAILS

A MODEL of Trevithick's third engine, the first locomotive in the world to run on rails

Modeled upon Franklin's example, electrical Parties of Pleasure sprang up everywhere. The Salem (Mass.) *Gazette* could advertise so on January 1, 1771:

> Tomorrow evening (if the air be dry) will be exhibited a course of experiments in that instructive and entertaining branch of natural philosophy called Electricity; to be accompanied by Methodical Lectures on the Nature and Properties of the wonderful element; by David Mason, at his dwelling-house near the North Bridge.

David Mason went Benjamin Franklin one better. Among other things he showed "fire darting from a lady's lips so that she may defy any person to salute her."

But all these fascinating expositions seemed to mark the final end of electrical discovery. They placed all the emphasis upon electricity as a toy. Experimentation into practical uses began to slow down. Franklin's kite, Pieter Van Musschenbroek's Leyden phial, the monk Gordon's bell and motor—all these were interesting, but what use were they or could they ever be?

In moments of discouragement not even the experimenters themselves could tell you. Jesse Ramsden had built his famous plate electrical machine, capable of giving more enthusiastic sparks than any glass cylinder. And Abraham Bennet, F.R.S., curate of Wirksworth, had developed in 1787 that intensely useful and interesting instrument, the gold-leaf electroscope. For all this, things were in the doldrums. Static electricity is, in the descriptions of to-day, too great in potential and too small in quantity to serve any useful purpose. It would have been no more effective, if put to work, than one microscopic drop of water, hurled at incredible speed, would have been effective to turn a water wheel. Electricity was only a rollicking colt in a world that had need of a pack horse.

And so, progress in the study of this mysterious electrical fluid all but came to a halt. It did not begin again with any enthusiasm until one day in the year 1786 an unimportant accident in the Bologna laboratory of an Italian physician who had his mind, not on electricity, but on the nervous system of the common frog, at one stroke opened a new trail.

CHAPTER VIII

The Slaughter of the Frogs

On a pleasant day in the spring of 1786 the thoughts of Dr. Luigi Galvani were far away from electricity. He was concentrating in his Bologna laboratory upon the nervous system of the frog. But Galvani's laboratory, like so many others, contained a frictional electric machine. And it came about that at the moment when Dr. Galvani's steel knife was probing into the minute muscles of a frog's skinned legs, its handle came into contact with a wire from the friction machine which stood on the same table. And it happened also that Dr. Galvani's assistant, at the same moment, gave the machine an idle twirl with his hand. And despite the fact that the frog was dead—so dead that all but its legs had been thrown away—these legs gave a vigorous and lifelike kick.

Dr. Galvani gave an equally vigorous and lifelike jump. Had he seen what he had seen? Was this a supernatural frog that he was so impiously dissecting? Could last night's bottle of Chianti have had anything to do with it? The doctor and his assistant were puzzled and a little frightened. They tried it again. Once again the legs of the dead frog did a delicate, macabre dance. After he had tried his experiment a dozen times more and the legs, with no show of fatigue, kept right on kicking, Dr. Galvani's fright wore off. Here, he decided, is a new philosophic discovery to be added to the world's treasured store.

That, at least, is one version of a story that has gone through as many disguises as the story of Cinderella. But even more curious—this was no discovery at all. The willingness of a dead frog's legs to kick under the right circumstances had been discovered a century before, but the world had forgotten it. John Swammerdam, a Dutch natural philosopher, had performed the experiment before the Grand Duke of Tuscany in 1678. It had never been published until, years after Swammerdam's death, his *Biblia Naturalis* came out in 1752. Even then no one seemed to have read these words:

> Let there be a cylindrical glass tube in the interior of which is placed a muscle, whence precedes a nerve that has been enveloped in its course with a small silver wire, so as to give us the power of raising it without pressing it too much or wounding it. This wire is made to pass through a ring bored in the extremity of a small copper support and soldered to a sort of piston or partition; but the little silver wire is so arranged that on passing between the glass and the piston, the nerve may be drawn by the hand and so touch the copper. The muscle is immediately seen to contract.

And in 1700, to add to the puzzle, Joseph du Verney had done something of just the same sort—preserved in the *Histoire de l'Académie des Sciences:*

> M. du Verney showed a frog, just dead, which, in taking the nerves of the belley that go to the thighs and legs, and irritating them with a little scalpel, trembled and suffered a sort of convulsion. Afterward, he cut the nerves, and, holding them a little stretched with his hand, he made them tremble again by the same motion of the scalpel.

Dr. Galvani, shivering with the joy of what he believed was a new discovery, felt quite positive that his lifelong search had been rewarded: that in electricity he had discovered the secret of all vitality. If electricity could create the semblance of life in the dead legs of a frog, then obviously, Galvani reasoned, electricity and life were one and the same thing. Quite neglecting the possible uses of his discovery, he went soaring off into the vague kingdom of metaphysics.

The immediate effect of Dr. Galvani's casual accident was the sudden death of about one hundred thousand frogs. The dismembered legs of these martyrs to science immediately began kicking in every laboratory in Europe, and the emotions of a mother frog on the approach of an eighteenth-century physicist were dark and despairing.

Dr. Galvani led the band of slaughterers. During the course of his studies he came, as Swammerdam and Du Verney had come before him, upon this further fascinating fact, the true significance of which he totally missed: when he hung a pair of legs by a copper hook against an iron grill, the legs, when the two metals came in contact, kicked just as joyously as they had when the frictional machine first set them off. Why was that? Quite unknown to Galvani, it was because there existed in the contact of two different metals the basic principle of the electric battery. But Galvani had another theory to prove—the theory that electricity was life. He explained this new kicking by pointing out that the frog had been freshly killed and that accordingly the life force of electricity had not completely seeped away.

Galvani succeeded in convincing almost every scientist in Europe by the book he wrote. Fortunately, one quizzical gentleman who taught natural philosophy at the University of Pavia remained resolutely unsure. He was Alessandro Giuseppe Antonio Anastasio Volta, and Galvani's proofs seemed to him too glib to be true. To Volta, electricity did not equal life. He said so, thus beginning a classic discussion that still goes on to-day and has not yet reached any exact conclusion largely because no one knows what anyone else means by the words involved in it.

Volta had a less ambitious explanation of why the legs kicked. They kicked because when they were hung by a copper hook against an iron grating something stimulated them from the outside just as the frictional machine had done. Volta had no theories to back him up . . . but what else could the stimulus come from except the combination of the two different metals? He stated quite firmly that this was a rational belief and defied the stout dictum of Dr. Galvani.

Galvani was promptly furious in a hearty, thorough-going Italian way. The two scientists began flinging insults at each other with a technique that no longshoreman could have bettered. Two dissimilar metals are not necessary, thundered Galvani, for the effect will take place with two pieces of the same metal. That, retorted Volta, purring between his clenched teeth, is an anomaly which is due to the differences in the temper or polish of the two pieces which causes them to act like different metals.

With this, one Dr. Valli rose to announce to the scientists that both were wrong and that he could make a frog's leg twitch by bringing two severed portions of the frog in contact and using no metal at all. That seemed to be a triumph for Galvani, who loudly renewed his insistences that electricity and the life force were indistinguishable and that Volta's idea that two separate metals had anything to do with the kicking of the frog's legs was entirely beside the point.

Here was a difficult matter for Volta to explain. He sulked in his laboratory for a while, quietly slaughtering frogs in search of the right answer. Soon, he decided, he had found it. He fired another salvo at Galvani, shattering what the latter thought was the tranquillity of victory. I notice, none the less, said Alessandro Volta, that the two parts of the frog must be unlike parts or else no twitch will occur. My metal contact theory was the truth but only part of a greater truth—which is that any combination of heterogeneous substances will produce the same effect. It was now Galvani's time for reflection. He fell silent but was not subdued, and fortified himself by galvanizing every slaughtered fragment of animal life upon which he could lay his trembling hands. I still know, he muttered in Volta's direction, that your theory is quite wrong.

Very well, said Alessandro Volta, I shall prove it to you. But before he could, Dr. Galvani made his final gesture of disdain. He died. It was almost two years later that Volta's proof for his seemingly shaky and baseless theory became generally known. Yet four years before that he began work by setting up a pile of metals. First a zinc disc. Then a piece of moist paper. Then a copper coin, in defiance of the majesty of the state. Then more moist paper. Then another zinc disc. And so on, an alternate pile of zinc and copper with wet paper sandwiched between until the pile had one hundred pairs of metal discs. To the top copper disc and the lower zinc disc Volta attached wires. He brought them close together, touched them to the legs of the frog, and was rewarded by a kick the like of which the world of zoölogy had seldom seen before.

Volta's pile of metal discs was not many inches high. But not in Stonehenge nor in the great Pyramid of Gizeh had man ever produced a higher monument. It dwarfed the Sphinx. It shrank the great stone Buddha of Tibet into insignificance.

For Alessandro Volta had been very right and Luigi Galvani had been very wrong. And out of Galvani's error and Volta's skepticism there had been born a force which, if not life itself, was destined to become, within one century, the single greatest adjunct to it. Out of this crooked, toppling cylinder of metal there had come the force of the first continuous electric current. Volta did not know why, he admitted, two different metals brought together resulted in an electrical flow. Yet if anyone doubted, Volta invited him to test the simple facts for himself. Let anyone who liked place a silver coin under his tongue. Let him then place a steel key on the top of his tongue and bring the shank of the key and the outer edge of the coin into contact. There is produced immediately a tingle and a sour taste as evidence that something was happening.

In Volta's earliest piles one hundred discs had been enough to produce a distinct shock, and five hundred would produce a disagreeable one. But Volta soon found that he could increase the electric flow still more if he moistened his sandwiched paper not with water but with sulphuric acid.

But here was a disadvantage; the pile was always drying out and refusing, when dry, to supply any flow at all. Volta knew that it was the acid, not the paper, which served the useful purpose. And so presently he displaced his pile by a "crown of cups." By this he gave the wet battery the form it has to-day. The crown of cups was merely a set of porcelain jars filled with acid and each containing a strip of silver and a strip of zinc. Connect twenty of them together, Volta found, in the cascade or series system of Dr. Franklin, and a current would flow strong enough to decompose water.

Beyond that, Volta had no immediately important uses for the electric current which he produced. Twitching the legs of a dead frog was excellent fun, but it did not advance the world's work. No more did his electrification of the head of a hanged criminal. He turned to other pursuits—tried making a cricket sing louder by the stimulus of electricity, and studied also to see if the flow from his pile would increase the luster of a beetle. Soon, as it always does, the flood of truth would begin to submerge the man whose observation had brought it forth; Sir Humphry Davy, in England, was already putting the electric flow to sensible uses and had begun the foundation of the great science of electro-chemistry.

In the same year that knowledge of Volta's pile began to

spread, Napoleon Bonaparte was at the full tide of his triumphs. In the midst of his brilliant swath across the world, he paused long enough in Paris on November 18, 1801, to attend the second of three lectures that Volta gave at the Institute of France. Napoleon was then First Consul and his mind was not on science. Yet when he saw this queer-looking Italian bend over his equally queer crown of cups and with it produce a current that tore atoms of hydrogen and oxygen apart in water, his swift-flashing enthusiasm sprang up. There was nothing in this invention for a military man, of course, but it was just as well to encourage natural philosophers. Napoleon proposed Volta for the Gold Medal of the Institute. A proposal from the First Consul was a command. Volta received the medal, with six thousand francs and the Cross of the Legion of Honor and the order of the Iron Crown thrown in. Having thus acted the part of the gracious patron to science, Napoleon left on the more important mission of making life miserable for Austria. The army that marched behind him was a small and unimportant body of men, measured by the power that was some day to be drawn from the pile which had made Alessandro Volta distinguished and rich.

Volta remained always in the good graces of Napoleon. In later years he was not well and considered retiring from his professorship.

> If Volta's functions fatigue him [said Napoleon] they must be reduced. Let him give only one lesson a year if he likes; but the University of Pavia will be struck to the heart the day I permit his name to disappear from its calendar. As for the rest, a good general ought to die on the field of honor.

And for his further glorification, the Emperor made him a count and a senator of Italy which, at the time, was under his management.

But even this could not save Volta from an attack of

apoplexy in 1823, or in 1826 from an exhausting bout with jaundice. On March 5, 1827, he died. But the seeds which he had sown were already being reaped at the time of his death at the gifted hands of Davy, Oersted, and Ampère. And his apparatus, which first produced in the world a tractable and tamable flow of electricity, rests now in the dignity it deserves, in the Palace of Brera, at Milan.

The idea had now reached men's minds that electricity could be a controllable force. Heretofore, the spark that leaped from a Leyden phial had been anything but docile. It went where it pleased, and as it pleased. It would punch a hole in a piece of paper or knock down soldiers of the Crown, and kill a chaffinch, a turkey, or a scientist. But like the wild asses of the devil, it would not be harnessed. Now came the voltaic pile to show that there was another kind of electricity, industrious and amenable to the will of man; a willing dray horse that would carry burdens.

Whereupon the stream of electricity joined another stream; a stream that had been flowing so sluggishly that even though it had risen before history began, it had scarcely, by Volta's time, got anywhere at all. The other stream went by the name of magnetism.

Despite the blight of the Middle Ages, Europe had known what a magnet was—knew so well that by the year 1450 it was using a magnetized needle floating on water as a compass. The Finns, who have often annoyed more southerly races by doing things before them, had used this same compass in the year 1000. Yet no one thought of Volta's pile and the mariner's compass as having anything in common. It is to the everlasting credit of Hans Christian Oersted that he blasted down a partition in men's minds which made out of electricity and magnetism two separate and disrelated things, instead of cousins so closely kin that men wonder now if they can be separated at all.

CHAPTER IX

Conspiracy of Genius

On August 14, 1777, the village apothecary in the small Danish town of Rudkjöbing, on the island of Langeland, closed his shop early and hurried nervously toward home. That day he became the father of his first child, who, after a proper interval, was named Hans Christian Oersted.

Thereafter, brothers and sisters followed fast, and since the income of a village apothecary in the late eighteenth century was not notably opulent, none of the children had much opportunity for schooling. But it was a simple day; the apothecaries of the eighteenth century ranked far higher in the scale of importance than do drug store keepers in the twentieth. Science was in no way organized as it is now, and apothecaries were not only dealers in drugs but were likewise chemists, physicists, and physicians. Oersted's father was interested in the "electric flow"; it was no more odd that an eighteenth-century apothecary should be so interested than that a modern one should be interested in selling alarm clocks. And so, despite his lack of schooling, Hans Christian Oersted early learned a few of the peculiarities of electricity.

By 1794 Hans had been tutored sufficiently well to enter the University of Copenhagen, where he began a six-year study of medicine and physics. When he graduated there seemed little else for him to do save fall back upon his early training and take charge of an apothecary shop in the great city of Copenhagen. But he had been a good scholar, and now he was able to augment his daily work with a series of lectures which he delivered on chemistry to the medical students of the university.

This was in the year 1800—the same one in which Alessandro Volta had invented the electric battery which had produced such a stir in scientific Europe. Oersted tried producing acids and alkalies by means of this new chemical weapon and succeeded none too well. He struck out bravely on two years of travel through Germany, France, and Holland, and in 1803 returned to Copenhagen where, with some temerity for a young man of twenty-six, he applied for the chair of physics at the university. The university decided that he would not do.

The townsfolk were more enthusiastic. They raised a grant of three hundred *bankthalers* per year from the public purse for three years and three hundred more to help him with his experiments. He was set up in a private laboratory now and gave lectures on natural philosophy. From his pen there flowed a series of scientific papers written in German and Danish. Nothing impresses the overseers of a university so much as a series of scientific contributions which they cannot understand. Oersted's had a magical effect upon the particular ones who pulled the strings at Copenhagen, and in 1806, at the age of twenty-nine, he became professor of natural philosophy at the University of Copenhagen. Thereupon he married his housekeeper and had three sons and four daughters of magnificent mediocrity.

Hans Christian Oersted was a simple and unassuming man. He was devoutly religious. Now and again he was mildly transported to the composition of poetry. He made the conventional gestures of defending the poor by helping indigent students. He had the correct views on the freedom of the press and was, perhaps, unorthodox in no way save in his insistence upon instituting a series of popular lectures on natural philosophy for the benefit of *young ladies*—this in the uncompromisingly masculine years that preceded 1820. He had long hair and a large nose and an extremely high forehead. Behind that forehead there burned, almost obscured by his soft manner and conventional ways, the authentic spark of genius.

Upon one subject he had an inner conviction for which no logic could account. He was quite certain that there was some sort of relationship between the electric flow of Volta's pile and the mysterious force which held a magnetic needle wavering toward the north. There was no evidence. The two phenomena were, on the surface, entirely disrelated. Yet Oersted was sure. He set out to make the needle respond to the pile. But he could not. Electricity was manifesting its character slowly, and for thirteen years Oersted pondered in his laboratory without being able to form any idea of how he might demonstrate its relationship to magnetism.

There came then the winter of 1819—the thirteenth year of Oersted's tenure in his chair of physics. He had, one day, a lecture to deliver privately to a handful of advanced students in natural philosophy who were considering the difficult matter of electricity. One of its peculiarities was already notable: it heated the wires through which it passed. For the delight of his class on this day Oersted had placed a fine platinum wire across the terminals of an electric battery and set about showing his students the way in which the battery's mysterious force caused the platinum wire to glow. On his desk there was a small compass which at one moment got in the way of his manipulations. He moved it across the desk under the wire—and stopped suddenly in his smooth professorial speech. He had seen something tremendous.

But he conquered his excitement and went on with his discourse. When at last it was over, his group of students unknotted themselves from about his lecture table and fled. In his empty laboratory, Oersted, his hand shaking and his heart racing with excitement, moved the compass once again beneath the charged electric wire. He had seen it right; the same tremendous thing happened again. The compass needle swung erratically from the north and pointed foolishly toward the Mediterranean Sea.

Oersted disconnected his battery wire and tried again. This time nothing happened. But whenever a current was flowing in the wire and Oersted brought the needle of the compass up to it in a parallel direction, the dancing compass needle moved suddenly on its pivot in complete defiance of all ideas of geography. And in its light gyrations lay all the future of modern electricity.

Oersted realized not one ten-thousandth part of what he had discovered—but this much, beyond the shadow of a doubt, he did know: an electric current had exactly the same effect upon a compass needle as had a magnet. His suspicion had been entirely right. Between electricity and magnetism there was a mysterious but vastly important relationship. He tried the effect of a bigger battery upon his compass needle, and its deflection from the north was even more unmistakable.

Here, then, was the greatest of all subjects for a scientific paper. Oersted sat down to labor over it, and by July 21, 1820, he had finished a tract in magnificent, orotund Latin, describing his experiment for the benefit of the learned societies of Europe; telling them that the flow of electricity in a wire produced a magnetic effect in a circular direction around that wire, such that a magnet tended to set itself at ninety degrees to the direction of the current and that if the current were reversed the needle of the magnet complied with some natural law by doing the same thing. The great science of electro-magnetism had begun. And before long, one of Oersted's greatest successors was to say of this apothecary's son that he had "opened the gates of a domain in science, dark till then, and filled it with a flood of light."

The account of this simple experiment, clothed in impenetrable Latin, set the scientists of Europe off into a frenzy of excitement. A vital clue had been picked up in the great philosophic riddle of the day, and honors and congratulations began to pour in upon the simple man who had found it. It had been an accident, true enough; most of the great discoveries and inventions in the world had been accidents before Oersted and would continue to be after him. It takes discernment to recognize a jewel in the rough. Oersted had done this; for it he was rewarded with the Copley Medal from the Royal Society in London and with three thousand francs from the more practically minded Institute of France.

The progress of science, and of the idea of power that lay behind it, was to begin now upon a tremendous acceleration. It had taken the modern world some twenty centuries to improve upon the Greeks; it was sixty years before Watt improved upon Newcomen; it was twenty years before Oersted improved upon Volta. But it was only one week before another man of natural philosophy improved upon Oersted. That man was André Marie Ampère.

He had been born in Lyons on January 22, 1775—some two years before the apothecary of Rudkjöbing had hurried home in time to welcome Hans Christian Oersted into the world. His father was a proud and prosperous merchant who retired soon after his son's birth and moved to the near-by village of Polemieux. There he studied his strange child, who sat on the floor and did sums by arranging pebbles and beans, and once so far forgot the elements of nutrition as to crumble a biscuit into pieces and study addition with the fragments of it.

So precocious a genius needed encouragement, his father thought, and shortly he took him back to Lyons again and gave him into the care of the Abbé Dubarron, librarian of the college. Here, after a little browsing, the young Ampère demanded two engineering works which would still tax the abilities of the average college student. He wished to be made privy to the beam and column theory of Euler and to the secrets of hydraulics as vouchsafed by the great engineer Bernouilli.

This seemed a little tough, even to the Abbé. "Bethink you!" he warned. "These books contain the differential calculus, and are among the most difficult that the human mind has produced." The young Ampère was sure of himself: "Nevertheless, I hope to be able to understand them," he replied, prim and thin-lipped. "But they are written in Latin!" said the Abbé softly. This had very little effect upon the determined twelve-year-old child, who forthwith began to learn the language.

The fall of the Bastille occurred when Ampère was fourteen. Afterwards, he listed it, with his First Communion and a eulogy on Descartes, among the three most prominent influences in his boyhood; at the time he spoke of the Bastille's fall as the finest indication of the nobility of man. He was to change his mind soon enough. His father, now administering justice in Lyons, in the midst of the Jacobin-Girondin strife, had sentenced the anarchist Chalier to death. Soon thereafter Lyons had been taken by the Army of the Convention and reprisals were visited. "Ampère the aristocrat" was hunted through the streets by a yelling mob, and on November 24, 1793, he was guillotined, one of the first victims of the Revolution now in full flower. He left a farewell letter to his wife saying, with that concern which is never far distant from the French heart, that his chief expense in life had been for books and mathematical instruments for their son André, but that, all in all, he regarded it as a wise economy. In that regard, the rest of the world was soon to agree with him.

Science Museum, London

"PUFFING BILLY"

THE noble product of early engineering genius in England

FIRST TRAIN

A MODEL of George Stephenson's product "The Locomotion," first steam locomotive on the Stockton & Darlington Railway

The shock of his father's death jarred the natural philosophy out of young Ampère's mind, leaving it empty and a little foolish. For the next three years he spent most of his time in aimless wanderings in the fields and gazings at the sky. Then, on an April evening in 1796, while aimlessly botanizing after sunset, he spied two girls on the bank of a brook. One of them was Julie Carron, and the sight of her seemed to restore his wandering intellect. He used an old device: from her family he borrowed books so that when he returned them he might meet her. He conveyed her to and from Mass, and by and by plucked up his courage to speak to her intimately of the science of trigonometry. Together they calculated the height of the village steeple. By July, 1797, he was so far lost to shame as to record in his diary that he had given her a lily. Then he held her hand in helping her over a garden wall. On August 2, 1799, the inevitable happened. They were married—and Ampère supported his new wife by giving private lessons in mathematics to students at Lyons. He was embarked on the career of a teacher now, and soon realized that academic preferment came only to those teachers who published scientific papers. In 1802 he applied mathematics to gambling and by his work, *Considération sur la théorie du jeu,* he demonstrated what no one will ever believe—that the odds are always against the winner. This work, apparently a deep consolation to some members of the National Institute, resulted in his appointment in 1803 to the chair of mathematics and astronomy at the newly founded Lycée of Lyons.

A year later his wife died; once again jarred from his foundations he went to Paris, remarried after a decent interval, and by the year 1809 had settled down comfortably in the chair of analytical calculus and mechanics at the Ecole Polytechnique. He was still there on September 11, 1820—the great day upon which the discovery of Hans Christian

Oersted was announced to the Institute. The week that followed was the great one of Ampère's life. The supple mind that had devoured the differential calculus at the age of twelve set to work, and on September 18th, Ampère astounded the Academy by presenting a theory of the relationship between electricity and magnetism far more clear and cogent than had been Oersted's own. Ampère was a brusque and excitable man, who usually worked erratically, but in those seven days he worked with such remorseless speed and accuracy that the wise men of the Academy not only laughed at his results but questioned his honesty. No man, they insisted, however brilliant, could perform so much original work in seven days.

Ampère showed the Academy that the deflections of Oersted's compass needle followed a definite rule. He showed the important fact that two parallel wires conveying current attracted each other when the current flowed in the same direction and repelled when it flowed in the opposite. He demonstrated that the force of the attraction or repulsion was directly proportional to the strength of the current and inversely proportional to the square of the distance between the two wires. It was for this the great James Clerk Maxwell, on that day still a child unborn, was to name his predecessor as the "Newton of electricity." Furthermore, said Ampère, and proved it too, a solenoid, or coil of wire, when a current is passed through it, behaves exactly like an iron magnet possessing north and south poles, and will, when free to swing, react to the earth's magnetic north exactly as a compass needle. As if these demonstrations were not enough, he further proved that the phenomenon of magnetism could be directly produced in a bar of soft iron if it were thrust into a solenoid in which a current was flowing. All in all, Ampère's week of intensive work was one of the greatest weeks in all the history of science; into it he compressed a lifetime of accomplishment.

He was a man who enjoyed an interminable game of chess, and was so long-winded a talker that he once lectured to a suffering friend for thirteen hours on the philosophy of the universe. He was one of the first of the "absent-minded professors" who were, for so many years thereafter, to provide staple amusement for the readers of comic magazines. Crossing the Pont des Arts one day, he paused to pick up and examine a pebble. Lost for some moments in geology, he recovered himself with a start, looked at his watch, saw that he was late, and began a mad dash for his classroom—not, however, before he had thrust the pebble tenderly into his pocket and thrown his watch into the river Seine. He died on June 10, 1836, of acute brain fever at Marseilles, and in 1844 the International Congress of Electricians immortalized his name by adopting the *ampere* as the unit to describe the rate of flow of an electric current.

Yet long before Ampère's death the science of electrical theory had flowed into even broader channels. There was still something of the medieval alchemist in the make-up of Alessandro Volta, who electrified corpses with his pile and sought to make crickets sing and beetles shine because of it. Galvani had touched off the mind of Volta; Volta had touched off Oersted; Oersted had touched off Ampère. Thereafter scientists engaged no longer in medieval speculations but in the cold, hard application of mathematics to a great new phenomenon.

One of the wisest and most unappreciated of these new men of science was Georg Simon Ohm, who had been born the eldest son of a master locksmith at Erlangen, Bavaria, in 1789, and who was thus a boy of eleven when Volta put his pile of copper and zinc together to produce the first steady electric flow. By 1817 Ohm was a teacher of mathematics in a Jesuit school at Cologne and had been impressed by the demonstration of the scientist Fourier, that the flow of heat in a metal was proportional to the difference in temperature at its two ends. Ohm had the generalizing mind. Heat was a mysterious quality about which chemists and physicists were still wrangling; still more mysterious was this virtue of electricity. Yet, so Ohm reasoned, more than likely there was some similarity in the actions of the two. Having gone thus far by analogy he constructed some crude apparatus of his own and very shortly had a firm grasp on the truth that the flow of an electric current in a wire depended directly on the difference in the potential (that is to say, the pressure, some day to be called the voltage) between its two ends. He further discovered that when this pressure was constant the flow of current varied inversely according to the resistance of the wire. And thus, in 1825–26 Schweigger's *Journal* and Poggendorff's *Annalen* carried the news of "Ohm's Law" to the scientific world—to the effect that the amount of current in a wire is equal to the electro-motive force of the battery divided by the resistance of the wire.

Here, once again, was a great scientific generalization, and one without which to-day there would be no such thing as the mathematics of electrical machinery. But it was altogether too much for the world of 1826, and no one chose to believe it. Ohm insisted upon its truth, and the minister of education, who had charge of seeing to it that young minds were not polluted by new ideas, intimated to Ohm that any physicist guilty of such baseless heresy was not wanted in a chair of physics in his dominion. The broken-hearted Ohm returned to Erlangen and for six years languished there, doing a little feeble and purposeless experimenting and wondering unhappily why anyone with a new idea was always treated as if he combined the worst features of a felon and a leper.

Yet his story had a happy ending. By 1833 the tide had turned, and Ohm was appointed by the Bavarian government to the chair of physics of the Polytechnic School of Nürnberg. In 1841 the Royal Society in London anointed "Ohm's Law" with its recognition, and the Copley Medal went to its hitherto repudiated discoverer.

Early in 1854 Ohm had an apoplectic stroke and in July (he had continued lecturing until the end) the "noiseless career" of this short, jolly, self-effacing man came to an end. Thirty years later the International Congress in Paris did for him what it did for Ampère and adopted the term *ohm* as the unit of resistance. It was a distinction not without its irony. Resistance had been not only Ohm's discovery but his fate.

Thus one river of science rose and gathered greatness as it flowed. The time was soon to come, now, when it was to be joined by another, and perhaps a greater one. The branch upon which Volta and Oersted, Ampère and Ohm had sailed rose in Italy; the other had its source in Woburn, Massachusetts, U. S. A.

There, on March, 26, 1753, one Benjamin Thompson was born. He was a precocious child who was intimately acquainted "in algebra, geometry, and even the higher mathematics." This arsenal of knowledge he brought to bear, when he was fourteen, upon the calculations of a solar eclipse and hit it within four seconds. But his precocity went even further than these useful subjects, for at nineteen he espoused a lady fourteen years his senior. "I married, or rather I was married," (he wrote) to the daughter of the Reverend Timothy Walker, who had settled in the minute town of Rumford, now Concord, New Hampshire.

He became a major at twenty-two, and in other ways was too gifted to be popular. When the American colonies revolted he fell under suspicion of his brother officers, who branded him a Royalist, and when he was sent to England as a despatch bearer he felt, quite pardonably, that it might be the Higher Patriotism not to return too soon.

In 1798 Benjamin Thompson was in London. He was Count Rumford of the Holy Roman Empire now, and the historian Gibbon had called him "Mr. Secretary-Colonel-Admiral-Philosopher Thompson" by way of indicating his numerous gifts and stations. In his capacity of philosopher he had devoted himself to many things, among them an inquiry into the reasons why all shivering Londoners had to put up with fireplaces which poured smoke lavishly into their rooms while all heat went up the chimneys. This, and the state of other practical devices in England, Count Rumford yearned to improve.

So it was that when his friend Sir Joseph Banks called a meeting which both attended, ostensibly for the purpose of "assisting the poor," Count Rumford forced the meeting far away from its agenda into a discussion of the need for an institution of the practical arts, which would deal not only with fireplaces, but with kitchen utensils, laundry apparatus, and models of "all such other machines and useful instruments as the managers of the institution shall deem worthy of public notice." He pressed for this and for a lecture room and laboratory for "teaching the application of science to the useful purposes of life." There were to be lectures upon heating and ventilating, the preservation of food, and the chemistry of such diverse operations as tanning, agriculture, digestion, bleaching, and dyeing. Thus it was that a charter was sealed on January 13, 1800, whereby there came into being an organization known as the Royal Institution, with the choleric but forceful Count Rumford as its guiding spirit.

Around him Count Rumford sought to create a circle of gifted young scientific men. Into his ken there swam a youthful Cornishman by the name of Humphry Davy, who soon became an assistant lecturer in chemistry. He received the sum of one hundred guineas per year, together with a room in the Institution and sufficient "coals and candles" to maintain him. It was Humphry Davy who discovered the great power of electricity upon chemicals, who invented the miners' safety lamp, and who discovered pleasant properties of laughing gas, which provided the society of the day with a new and fashionable means of intoxication. Soon this young Cornishman became Sir Humphry Davy, Bart., and the discoveries which poured from his fertile mind gave a new fascinating impetus to the sciences of chemistry and physics. Yet his greatest discovery was, perhaps, the discovery of the mind which lay hidden in the unpromising skull of a blacksmith's son, who had once the temerity to apply to him for the humble work of washing bottles. With that, a whole new art began.

CHAPTER X

THE COACHMAN'S KNOCK

ON THE twenty-second of September, 1791, just eight years before Count Rumford had brought the Royal Institution into being, a blacksmith, whose forge was one of the industries of the village of Newington Butts, in Surrey, breathed a sigh of relief and wiped the sweat from his forehead on the news that his wife Margaret had successfully brought into the world his third child, a son. The blacksmith's name was James Faraday, and the name upon which he and his wife soon fixed for their weeping infant was Michael.

He grew up within the cramped and unhappy atmosphere of a poor laborer's household, without horizon, without prospects, without play, and without schooling. There were no compulsory education laws to apply to a boy who was seven years old in the year 1798, and in this year young Michael Faraday was put out by his hard-pressed father and mother to earn what money he could as a newsboy in the stern, economical day in which periodical publications were rented, not sold. It was at this delicate age that the philosophical streak first showed in him. He put his head, one day, through the bars of an iron fence so that his ears projected just beyond it. "Now," he asked himself, earnestly, "which side of the fence am I on?" Even to-day there is no very satisfactory answer to that question. Michael Faraday got nowhere with his first scientific experiment for the reason that the owner of the fence took prompt and efficient means to prevent its defilement by a ragged messenger boy. He picked Michael out from between the bars and set him very violently and uncontrovertibly upon the outside of the fence. That was the end of that experiment.

In 1804, when he was thirteen, Michael Faraday had a rise in the world of commerce. One Mr. George Riebau, a bookseller and stationer, Number Two, Blandford Street, London, agreed to take the young Michael on a twelve months' trial as an errand boy, with the agreement that if he displayed sufficient ability in that exacting business he would then be glorified by a formal apprenticeship to learn the trade of bookbinding. Michael stood the test, and one evening in 1805 he bounded home exultantly to tell his family that success was his and that a great career stretched endlessly before him. He was a bookbinder's apprentice! Michael began briskly upon this career. But early in the process of his apprenticeship he developed a vicious habit, not to be encouraged among the members of his trade—something that slowed up production and subverted discipline. He began to read the books he was supposed to bind. Now and again Mr. Riebau would discover him at his malingering and, although a kindly man, he could scarcely be expected to brook this violation of indentures.

By one of those odd turns for which biology is famous, but not too famous, the mind of Michael Faraday presented a rich soil for ideas. The practice of reading developed for him a fascinating world about which he yearned to know. Somehow, learning in "natural philosophy" took a deep hold upon him. One day, on a walk near Fleet Street, he saw a bill upon the hoardings which unaccountably stirred his heart. It stated that a Mr. Tatum would shortly begin to deliver a series of lectures upon various subjects in the domain of natural philosophy, once a week at 53 Dorset Street, Salisbury Square, admission one shilling.

This great experience came to Michael at about the age

of eighteen. His sudden longing to hear Mr. Tatum lecture was so intense that even his brother Robert, who was then carrying on the family tradition of blacksmithing, was impressed and volunteered out of the family earnings to present Michael with enough shilling pieces to get him into most of Mr. Tatum's lectures. With this unexpected encouragement, Michael attended twelve or thirteen of them between February, 1810, and September, 1811. Sitting well forward on the edge of a bench, he drank in this fascinating new information of the existence of which shortly before he had had not the faintest suspicion.

Michael Faraday knew no mathematics. Fortunately for him and for the future of science, natural philosophy then made little use of that more patrician branch of learning. Fortunately, also, scientific experiment in electricity was a simple matter; Alessandro Volta had displayed his voltaic pile only twelve years before, and anyone who could accumulate a few discs of zinc and copper could make one and study its effects.

That was one of the first things that Michael Faraday did. Using a halfpenny as his pattern, he cut out seven discs of zinc and then, with a complete disregard for expense, he sandwiched them between seven ha'pennies. Between the layers he inserted paper soaked in concentrated salt solution and proceeded then to verify for himself the experiments which the great Volta had already recorded. Therein Michael Faraday, the blacksmith's boy, with less opportunity to understand scientific method and procedure than any newsboy seen selling papers on the street to-day, established one of his primary characteristics. He would make no assumptions, he would jump to no conclusions, he would take no one's word for anything until he saw the facts verified for himself. "I could trust a fact," he wrote in his later life, "and always cross-examined an assertion." He was no more than nineteen years old when he cross-examined his first major assertion.

The assertion stood the test. The fact that an electric current from a voltaic pile would decompose certain chemicals in solution was the great scientific novelty of Faraday's adolescence. No one knew why, but everyone was delighted to see it happen. Michael was thoroughly happy when, with his first battery, he decomposed magnesium sulphate (which in a more vulgar tongue is known as Epsom salts) and later when with a larger pile he decomposed copper sulphate and that more complex organic chemical, lead acetate.

Following which there came the great year of 1812. The map of Europe was in process of violent alteration. But during the time that portentous changes were taking place upon the feverish continent of Europe, the British Isles were the theater of a great scientific meeting. Napoleon and Wellington are dead, and their work died with them. Equally dead are Sir Humphry Davy and Michael Faraday. But on February 29, March 14, April 8, and April 10, 1812, the minds of these two men met with a thunderous detonation not to be equaled upon any battlefield. It was upon those four dates that the twenty-one-year-old Michael Faraday sat in the presence of the director of the Royal Institution and drank in the great and time-enduring dicta of natural philosophy.

Very carefully, very minutely, Michael Faraday took notes upon the lectures which Humphry Davy had uttered in the bare, echoing lecture hall of the Institution's building upon Albemarle Street. At their conclusion Michael Faraday knew that his future did not lie in bookbinding. Taking some of his notes, including minute drawings of the apparatus, Faraday wrote to Sir Joseph Banks, co-founder of the Institution. He would like, Michael said, to be employed in some fashion.

There was no answer to this letter.

Faraday winced under the rebuff. But by the end of 1812

he had again become too restless to endure inaction any longer. Trade seemed vicious and selfish to him; he yearned to experience what he then believed to be "the superior moral feelings of philosophic men." And so one evening, with a trembling earnestness, he sat down and addressed a letter, this time to Sir Humphry Davy, timidly suggesting that if there were any menial tasks to be performed about the laboratory he would be only too happy to care for them if he might in return have a few shillings upon which to live. With his letter Faraday once again enclosed some lecture notes as "proof of my earnestness." It is to the eternal credit of Sir Humphry Davy that, having broken the seals of Faraday's letter with a patrician hand and stared haughtily at the neat gentility of its script, he saw in it a spirit and a mind that he could use. To a Mr. Pepys, Sir Humphry confided that he had had an affecting letter from the young son of a blacksmith who was ambitious to study natural philosophy. He expressed a wonder about what it would be wise to do.

"Do?" said Pepys. "Put him to wash bottles; if he is good for anything he will do it directly; if he refuses he is good for nothing." And so it was that Sir Humphry, a day before Christmas, wrote out a reply, signed it with a brave flourish of his quill, dusted it with sand, and sealed it with a swab of wax. This is what it said:

Mr. P. Faraday, 188 Weymouth St., Portland Place

December 24, 1812

SIR,—I am far from displeased with the proof you have given me of your confidence, and which displays great zeal, power of memory, and attention. I am obliged to go out of Town, and shall not be settled in town till the end of Jany. I will then see you at any time you wish. It would gratify me to be of any service to you; I wish it may be in my power.

I am Sir
Your obt. humble servt.
H. DAVY"

In a shining joy "Mr. P. Faraday" awaited eagerly the coming of that Jany which would see Sir Humphry settled in town. It eventually came.

The two met in the anteroom to the lecture hall of the Royal Institution and Faraday, very frightened and white of face, hideously conscious that his clothes did not fit and that his shoes were in need of varnish, stood trembling at the temerity of his own voice which somehow, with no conscious bidding of his own, was telling Sir Humphry Davy that he wished to enter the service of science by the washing of bottles and other glassware.

Sir Humphry was prepared for a certain lack of polish in his applicant but hardly for the crudity of this untaught blacksmith's son. He saw the earnestness that showed from under the dusty clothes, but for all of the "superior moral feelings of philosophic men" he could scarcely bring himself to believe that Faraday would do. He advised him to stick to bookbinding, and the heart-broken Faraday stumbled out of the anteroom with his only door of opportunity apparently closed fast behind him.

Sir Humphry, dismissing the incident from his mind, went back to his researches.

But it was only a few weeks later that these researches brought him to the examination of that skittish compound called nitrogen trichloride, which had been discovered just two years before and which is still distinguished by the peculiarity that no one yet knows just why it explodes, or what becomes of it when it does. But there is no doubt that it does explode, and in one of its more temperamental moments a good quantity of it blew up in Sir Humphry Davy's face and temporarily blinded him in one eye. In the attempt to continue his work, Davy bethought him of the earnest Faraday, who might be a lout but who at least wrote a neat hand.

So it was that the shabby Faraday household on Weymouth Street was startled out of its painful gentility one night during March, 1813, when, with many flourishes and much clanking of heavy ornamented harness, a grand coach drew up to the door and an equally grand coachman alighted and knocked loudly upon its frailty. It was that coachman's knock that summoned Michael Faraday away from the trade of bookbinding and into a scientific immortality and which summoned the world out of its old ways of living.

As is so often true with all of us, the twenty-two-year-old Faraday was in a slightly awkward position to answer opportunity's knock. He was retiring for the evening and had just removed his trousers. But when he broke the seal and read the letter which the coachman handed him, his desire for sleep vanished altogether. Sir Humphry Davy, the great Sir Humphry Davy, director of the Royal Institution, was expecting Michael Faraday to call upon him once again—on the following morning.

So well, on second acquaintance, did Faraday's capabilities as an amanuensis impress Sir Humphry, that shortly Faraday found himself no longer a bookbinder's apprentice but a laboratory assistant to the director of the Royal Institution with a salary of twenty-five shillings a week, and two rooms at the top of the Institution's home in which to live. His horizon broadened swiftly, for in October, 1813, he left with Davy for the continent and traveled there with him until April, 1815, as an assistant during the course of lectures which Davy presented at such centers of learning as Paris, Lyons, Nice, Rome, Naples, Genoa, and Florence. Faraday had never before been farther away from London than twelve miles. He came home with the knowledge of a new world of strange, fascinating customs and a great feeling of scientific brotherhood with all the world. It is true his life was not an unmixed happiness. Lady Davy had accompanied her husband and quite failed to see the affection which he was beginning to feel for his clumsy young assistant whose hair did not comb readily and whose manicure was sometimes less than immaculate. She insisted that he was a servant and accordingly the future Copley Medalist ate his meals in the scullery.

By 1816 definite promise had begun to show in him. In that year he made a simple and easy analysis of caustic lime, and thus began his own career of original research. This began shortly to blossom to such a degree that by May, 1821, Faraday had become "Superintendent of the House and Laboratory" for the Royal Institution at a salary of £100 a year. Here was not only fame but riches. Straightway, on June 12, 1821, Faraday married a Miss Sarah Barnard, and with that detail out of the way, he settled down, a sober and established young man, to the serious business of natural philosophy.

One year before, Hans Christian Oersted had discovered that the magnetic needle of a compass might be deflected if current passed over it through a wire. Oersted's discovery was just as spectacular in 1820 as would be the demonstration to-morrow that there was a direct relationship between violin-playing and appendicitis. Oersted summed up his observation by saying that "the electrical conflict acts in a revolving manner" around any wire through which a current is flowing.

Most other scientists, immersed in attraction and repulsion, forgot that the "conflict acts in a revolving manner." Faraday did not. This idea of revolution seemed to him much more important than finding out why electricity and magnetism hypnotized iron filings in right-angled ways. Oersted had shown that there was some very real but rather confusing relationship between electricity and magnetism. Faraday set out grimly to discover just what it was. And if the electrical conflict acted in a revolving manner, Faraday would like to find out just what revolved about what. To that end he constructed a simple apparatus made out of a bar magnet, with its north pole upright, held vertical by wax to the base of a vase filled with mercury, and a copper wire which hung above the vase, and dipped slightly below the surface of the mercury in the middle of which the magnet projected. Out of this apparatus Faraday made a circuit by connecting the wire and the mercury to the poles of a voltaic battery.

All this machinery was more complex than any electrical contrivance of previous years. Faraday had the hopeful idea that when he completed the circuit and let an electric current flow through the wire and the mercury something would revolve about something else. On September 3, 1821, there came the day on which he was ready to toss his one greatest scientific coin and abide by the result. His apparatus was assembled; he had only to close one switch to see whether he was on the track of a great idea or was merely being rather silly in thinking that he knew more than his elders. He waited for a breathless instant—and closed the switch.

Whereupon the electric wire, its free end submerged in the pool of mercury, began a happy but sedate circle around the pole of the inscrutable magnet, and Oersted's theory had not only a verification but an extension into one of the single most important and far-reaching scientific observations that any one man has ever made.

That was enough for one day. Faraday, sober, devout, Sandemanian Presbyterian that he was, could not restrain his joy. He clapped his hands together and gave a wild caper about the table that almost upset the entire apparatus. "There they go! There they go! We have succeeded at last!" he cried. Whereupon he gave up science for the day and took the evening off to visit Astley's Circus and marvel at the

Science Museum, London

BLENKINSOP'S GEARED LOCOMOTIVE

A MODEL of the conscientious but mistaken engine which might have throttled railroading in its cradle

Smithsonian Institution

"THE ROCKET"

A REPRODUCTION of the famous design by George and Robert Stephenson which inaugurated practical railroading in England

ability of its performing horses. The horses performed quite unemotionally, not knowing that there sat in the audience a man who had that day taken from their shoulders the burden of the major work of the world and that their great great-grandsons need never suffer again as had suffered the horses on the treadmills of Cornish mines for all the years gone by.

Many a man of science would have rested his fame upon this one discovery. But Faraday went even further, and soon he had set up three great electrical commandments upon which hung, in the words of the New Testament, all the Law and the Prophets. They were these:

Under the influence of electricity a wire could be made to rotate around the pole of a magnet.

Conversely, if the wire were fixed, the pole of a magnet, if free, could be made to rotate about the wire.

Regardless of which rotated about which, the direction of rotation would be reversed if the connections of the circuit to the battery were themselves reversed.

In those three basic principles which Michael Faraday discovered within a comparatively few days of 1821 and 1822, there lay enough hard, rich, scientific nuggets of thought to keep him mining industriously during the next ten years.

It was quite soon obvious to Sir Humphry Davy that the awkward blacksmith's boy, who had come to him eight years ago and who had barely turned thirty, was a formidable scientific rival whose extraordinary mind was all too apt to eclipse his own. And so, among other difficulties, Michael Faraday had to contend with the bitter jealousy of the man who had once been his mentor and friend. Davy's jealousy reached such a pitch that he bitterly opposed the suggestion that Faraday, for his electromagnetic discoveries, be elected a fellow of the Royal Society. But Davy's health had begun to fail, and the man of whom Coleridge had said "Had he not been the first chemist he would have been the first poet of his age" was becoming mean-spirited and quarrelsome. In spite of his objections, Faraday was elected a fellow of the Royal Society on January 8, 1824, and, a little more than a year later, Davy's health having made necessary his resignation, Faraday supplanted him as "Director of the Laboratory of the Royal Institution of Great Britain." Although he was supposed to be Scientific Adviser to the Admiralty, at £200 per year, he never drew his salary and once even rebuked the Admiralty (anyone who thinks that is a small achievement should try it) for asking his opinion on someone's patented disinfecting powder. Neither Oxford's proudest award, the Doctorate of Civil Law, nor the Royal Society's highest gift, the Copley Medal, produced in Faraday the slightest desire to live the great social or political life that was now open to him, or to earn any salary more magnificent than one which would not now support a messenger boy. He continued to draw his £100 a year and live with his wife in the two rooms set apart for his use by the Royal Institution.

Thanks to the combined wisdom of Oersted and Faraday there now existed a great link between those hitherto separate things, electricity and magnetism. What Faraday had done had been to take a force, rampant, unmanageable, and ill-tempered, and produce out of it calm, leisurely, and ordered motion.

But once again Faraday went further. If electricity would produce motion, was it not also possible that motion would produce electricity? Gilbert had already answered that question favorably with his rotating sulphur ball, but what Faraday wanted to produce was not the visual spark of a friction machine but something akin to the even current that flowed from one of Volta's cells. Moreover, Faraday believed that if magnetism could induce magnetism it was quite possible that a current might induce a current—that is, that an electric flow could somehow be produced in one wire by a flow through another not connected to it. Faraday lay two wires sides by side, one connected to a voltaic pile, the other to a convenient instrument known as a galvanometer (ancestor of present-day volt-meters and ammeters) but could not, still lacking an important secret, shake the galvanometer needle out of its imperturbable calm.

For six years the secret eluded him. Then, in 1831, the proper relationship of circuits suddenly dawned in his mind. Around a ring of soft iron he wound two separate coils of wire. One coil he placed in circuit with a battery; the other with a galvanometer by which the amount of the current could be measured. On August 29th he stood ready for a new test. He closed the switch connecting one coil to the battery—and on the instant the needle of the galvanometer connected to the other coil jumped and was at rest again. Faraday hung upon the brink of disappointment. That jump was heartening and significant, but why was it no more than a jump? Why did a *steady* current not flow in the galvanometer coil during all the time that it flowed in the battery coil? Sadly, Faraday opened the battery switch—and was once again rewarded by a mysterious jump of the needle. Disappointment became elation; every time the battery switch was closed or opened something happened to the galvanometer needle. Very well; it was obvious that induction depended upon something more than mere propinquity of coils. It depended also on a starting or stopping of one electrical flow.

Faraday was close to his greatest discovery now. What he had almost unknowingly created was the device now known as a "transformer," but not until alternating currents became known would the exact nature of his discovery become evident. Yet Faraday now knew enough; the way had been opened to his most spectacular development—the long-sought-for production of an electrical flow from the force of magnetism.

Some five days went by while Faraday pondered and puzzled over this question and carried out many another experiment in his little stone-flagged basement laboratory at Number 20, Albemarle Street, or gave lectures in the tiered hall above. Faraday made a coil out of 220 feet of copper wire and mounted it in a wooden frame. Up to the hole in the coil he brought the end of a cylindrical bar magnet, a dense piece of black iron three quarters of an inch in diameter and eight and one half inches long. The ends of the copper wire he connected to his galvanometer, and then *very quickly* he thrust the magnet for its full length through the center of his copper coil. The galvanometer needle jumped, a mere eyelash of a jump. But it had indubitably moved as the result of the quick motion the iron bar had made through the vacant center of the copper coil.

Motion—quick motion—that was it. One might place a magnet and a coil of wire side by side and wait for a thousand years and nothing would happen. But move one of them sharply, relative to the other, and immediately the mysterious force of electricity coursed through the copper veins of the coil. He had made one more great generalization. Oersted had shown that there was a relationship between electricity and magnetism. Faraday now knew precisely what it was and how, starting with one, the other might be produced. Start with electricity and end up with magnetism and you have a motor. Start with magnetism and end up with electricity and you have a dynamo. Faraday decided (and his theories have lasted even into this present complicated day) that electricity is generated when a conductor is set to work "cutting the magnetic curves" which Peter Musschenbroek had mapped around the poles of a magnet from the way iron filings shaped themselves in its presence.

The next great thing was to transform this feeble current, which would give only a temporary flick to the galvanometer needle, into something that was steady and continuous—that is, to devise some sort of rotating machine, such that a continuous turning of a crank would produce an equally continuous deflection of the galvanometer needle. His coil and bar magnet were no good for that purpose. When the bar was thrust in, the current moved one way; when it was pulled out the galvanometer needle swung the other way, indicating that the flow of current was reversed. But suppose, he wondered, suppose I keep my magnet still and substitute for the coil of wire a copper disc mounted on a brass axle to be turned by a crank so that the disc will rotate between the poles of the magnet? Then, as the disc cuts the invisible lines of force, a current should be generated in it and, if I put one collecting brush on the brim of the disc and another on the axle, I may be able to tap and draw off some of the mysteriously invisible fluid in the disc.

Faraday built his machine out of a disc of copper twelve inches in diameter and a fifth of an inch thick, mounted it on an axle, and set it within the narrow space that separated the poles of a magnet—the great magnet of the Royal Society. From the collecting brushes he ran wires to his galvanometer and turned the heavy disc. This time the galvanometer needle did not flick. Before the eyes of the Royal Society's fellows, straining in the musty gloom of a lamplit London night, the needle swung over and hung steadily in one position, to show Faraday quite certainly that he was doing something that no man had ever done before; out of mechanical motion he was producing a tractable and presumably useful flow of electrical current.

To carry his induction theory further, Faraday performed his famous "ice-pail experiment" in December, 1835. He insulated a metal ice pail and connected it to a gold leaf electroscope on the bench beside it. Then by means of a frictional machine he charged a piece of metal positively. But immediately when he lowered this conductor, by a silk thread, into the ice pail, the electroscope to which the ice pail was connected promptly spread its leaves apart. So long as the conductor was not brought too close to the top of the can, it made no difference to the leaves what position inside the can the conductor might have. When the metal was removed, the leaves collapsed. Then, said Faraday, I have obviously *induced* a charge on the can equal and opposite to the charge on the conductor.

This fundamental and essential idea that the sum total of electrical charges was always zero had lurked in men's minds for a long time before. But not until Faraday and his ice pail, plus his calm, simple logic of the kind that makes everyone else wonder why he, himself, had not thought of it before, did the notion get itself demonstrated beyond the chance of being lost or contradicted again. You rubbed amber with silk and charged the amber one way. But, at the very same time, you charged the silk another way. In other words, you disturbed the simple, ordinary state of things by giving the amber too much of Something and the silk too little of Something—or the other way around; it did not matter much which. This idea later proved so necessary to all intelligent thought about physics or chemistry that these sciences could hardly have progressed very far beyond the phlogiston theory without it.

But Faraday needed one more linkage in his proof. In January, 1836, he built a box twelve feet on a side—large enough so that he himself could enter it—and coated it outside with tinfoil. After the fashion of the ice pail he insulated it and connected it to an electroscope. Then he went inside the box, lugging a frictional machine. He set it up on its bearings and ground away at it for fifteen minutes—until its bearings almost smoked and the air was heavy with the pungent smell of ozone. Then he stole a glance at his electroscope. Its gold leaves lay limp and flaccid inside the little glass bulb. That proves half my point, thought Faraday. Half my machine has a violent charge of one sort but the other half has just as violent a charge of the other sort. They add up to nothing.

Even now the proof was not complete. Suppose he put the electroscope inside the box and the frictional machine outside—what would happen then? He did, and although he charged the box until it spat sparks six inches long, the electroscope preserved the same calm, imperturbable lack of interest that it had shown before. The proof was complete.

Thereafter men knew one vital thing about the electric flow and what caused it that they had never been sure of before. From now on it was quite unnecessary to attribute it to auras, spirits, demons, or the other conveniences of the medieval world which talked, and talked, and talked, and never knew what it was talking about.

Michael Faraday was a simple soul. He was a devout Sandemanian Presbyterian, never troubled by the necessity of reconciling scientific discovery with the dogmas of the religion he cherished. And if he was once in high glee at being fortunate enough to see a bolt of lightning strike a church steeple in Eastbourne, it is hard to know whether it was the glee of a scientific man observing an exciting natural phenomenon, or the glee of a Sandemanian watching the hand of God smite a temple of unorthodox worship. He enjoyed watching the animals of the zoölogical gardens, found an unfailing delight in a Punch and Judy show, and rejoiced particularly in the Christmas lectures for children which for many years he delivered in the high drafty lecture room of the Royal Institution.

Upon these, and all his lectures, he lavished a care and pains known to few of the complicated scientists of this century. He studied elocution, for the greater pleasure of his audiences, never repeated a phrase, never amended a sentence once he had uttered it, propped a card with the word "Slow" written on it in front of him that he might not talk too fast for the audience to follow, and never lectured for more than sixty minutes by the clock. Wood, paper, wire, turnips, and potatoes all played their parts with him in the construction of models to illustrate obscure points of his discourse. He once made an electrical friction machine out of a quart bottle mounted on an inverted stool, with a tea canister on a tumbler for a conductor and a toasting fork for the collector. And despite the fact that he once wrote, "Lectures that really teach will never be popular; lectures which are popular will never really teach," he managed to achieve an extraordinarily fine blend of popularity with instruction.

He had the typical social restraint of a sensitive man and seldom appeared anywhere, save to answer the command of royalty. His personal participation in the arts was confined to the simple and comparatively inoffensive practice of playing upon the flute. He denounced the charlatanry of table-turning and spirit-rapping, and for his pains earned in return the blistering denunciation of Elizabeth Barrett Browning, who poured out her scorn upon him as a "shallow materialist."

Upon not one of his inventions did Michael Faraday ever apply for a patent. He declined knighthood from her most gracious majesty Queen Victoria, and when, in 1857, he was offered that most illustrious of all scientific honors, the presidency of the Royal Society, he wrote to John Tyndall, later to succeed him at the Royal Institution:

> I must remain plain Michael Faraday to the last; and let me now tell you, that if I accepted the honor which the Royal

Society desires to confer upon me, I could not answer for the integrity of my intellect for a single year.

By 1849 the mind so brimming with imagination and so closely crammed with fact had begun to fail. He performed experiments, only to discover that he had done them before. The well-springs of theorizing slowly dried. At Christmas time, 1860, he delivered his last juvenile lecture, and his last Friday Night Discourse he made before the Institution on June 20, 1862. His famous notebooks, in which he painstakingly recorded every experiment he performed and every scientific thought that crossed his mind, he had closed forever a few months earlier, on March 12th, with notation number 16,041—having to do with the effect of a beam of light upon a magnetic field. It was something not to be discovered definitely for another thirty-five years.

On the 26th of August, 1867, he died, peacefully sitting in the chair of his study. Four days later, without a shred of pomp, he was buried in Highgate Cemetery. The blacksmith's boy who was never certain of his arithmetic had been twice a Copley Medalist, and had died with a total of ninety-seven diplomas stored in his garret. There, too, his administrators discovered an old box. They opened it, and into a disdainful heap at their feet there poured the host of medals which every scientific institution in Europe had showered upon him.

CHAPTER XI

THE IRON COLT

FOR some time prior to the year 1763 the fellow artillery officers of Nicholas Joseph Cugnot had held the suspicion that something was wrong with him. Something was. He had an idea. He was being teased and tormented by the notion, which he scarcely dared to utter, that the steam engine might be applied to a moving vehicle.

Now this was absurd. The idea of steam locomotion had occurred to so great a man as Sir Isaac Newton, who had thought in 1680 to propel a carriage by steam. His idea was to mount a boiler upon a pair of wheels and drive the carriage by reaction from a jet of steam, spurting from a nozzle at the rear of the boiler. But Newton had been sensible enough not to put his invention to the test. If Newton realized his folly, why should Cugnot hope to accomplish anything with no better motive power than the world of 1763 afforded?

He persisted in his dream, and in the year 1763 he completed the construction of his crazy machine. Few details of it remain, but it is quite certain that in some measure it must have worked—for out of its performance there came sufficient governmental enthusiasm not only to permit but to encourage Cugnot to go further.

Meanwhile, James Watt had invented the separate condenser and out of Newcomen's queasy steam bellows had begun to create a real machine. Not even Watt would subscribe to the idea that a steam engine could be empowered with locomotion. In Watt's mind a steam engine was a steam engine and a stagecoach was a stagecoach. One did not belong in the domain of another.

Cugnot went steadfastly ahead with his designs. By 1769 he had completed a second steam-driven cart. That strange beast, the centaur, was no stranger anatomically than this. The centaur's rear quarters are a horse and its fore quarters are a man, but the rear quarters of Cugnot's machine were a wagon of heavy frame and thick, wobbling wheels, whereas the fore quarters seemed to be a combination of a whiskey still with a wheelbarrow. At the extreme front of the machine was a large, glistening copper boiler connected by a pipe to two thirteen-inch cylinders just over the single front wheel. The pistons of the cylinders were fastened to the axle of the wheel by a combination of connecting rod, pawl, and ratchet.

Although Cugnot's invention—which came in English to be called a steam cart—was a horrible-looking machine, it entranced General Gribeauval, inspector general of artillery, and the Duc de Choiseul, foreign minister and minister of war for Louis XV. The Duc de Choiseul was not an easy man to entrance. He was one of the most violent men in the violent years that preceded the French Revolution. But when at the Paris arsenal he saw Cugnot's steam cart muster its strength sufficiently to carry four men for a brief, thunderous, and horribly uncomfortable ride, he became as a little child, embraced Cugnot warmly, and bade him godspeed in his further ventures.

Perhaps it was this which encouraged Cugnot to the lamentable rashness of giving his cart a public trial. Some time later he took it into the streets of Paris and began a brief voyage of discovery. The stability of it, as may readily be appreciated from the illustration facing page 94, was not too good. The single front wheel was used for steering as well as propulsion. After a few moments of travel, which threw the volatile spectators into an acute panic, Cugnot attempted the difficult business of turning a corner. The unbalanced rear wheels rose for a moment and waggled weakly in the air—whereupon the entire machine fell heavily upon its side. At this crime French justice, as soon as it recovered its aplomb, struck hard and swift. Cugnot was flung into jail. The steam cart was hauled back to the arsenal and left to gather dust until, in later, more appreciative years, it was transferred to the Conservatoire Nationale des Arts et Métiers in Paris. There, it still proudly but dementedly resides.

In England James Watt, no doubt, took this mishap as a vindication of his disbelief in travel by steam, but his assistant, William Murdoch, did nothing of the sort. With Cugnot he felt certain that he could give the steam engine mobility. He built a model, fifteen inches high, of a steam cart. It was a small steam engine of walking-beam construction with a two-inch stroke and a ¾-inch diameter cylinder, mounted upon a little wagon with rear wheels 9½ inches high. Then one day he took it out to a country road for test.

Wisely, Mr. Murdoch sought solitude. He wanted to be neither jeered, stoned, nor jailed. He chose an hour of twilight in a road passing a cemetery, and thither he went with his bulky little model under his arm. There he glanced about furtively, set his model down, and stoked a miniature fire under its boiler. Soon it had a head of steam, and Murdoch gave it a cautious push. To his vast delight it began a noisy waddle down the road. Murdoch stalked beside it, bursting with the pride of a father whose son has just learned to walk.

This was too much for the fates. Thus it turned out that the vicar of the parish whose dead were resting close to Mr. Murdoch's desecration chose that same road for a brief evening stroll, his head filled with the vague but comforting thoughts of vicars. As he approached the graveyard there burst upon him the ghastly scene of a spectre walking beside some device clearly possessed of the devil, since, although inanimate, it was capable of that sole prerogative of life which was locomotion. The vicar was not merely horrified; he was aghast.

There was a confused interval in which Mr. Murdoch sought to calm the vicar, whose nerves as well as religious beliefs had been given a wrench. But as the former waned the latter waxed, and once Mr. Murdoch had made it quite clear that here was no instrument of Apollyon, but merely a phenomenon of natural philosophy trying to express itself, the vicar resumed his command of the situation and settled down to an impromptu but eloquent sermon. It has not been recorded, but it fixed Mr. Murdoch. He pleaded forgiveness and went home and abandoned his model. The vicar prayed long that night, and steam transportation rejoined the company of hell, whence it came.

Thus the business gentlemen who had need to travel from London to Edinburgh still required some weeks by stagecoach. Nevertheless there were forces at work which were to change matters.

By great good fortune a Cornish mine engineer, who knew what the problems of heavy transportation were, had had some experience with Murdoch at the time he was working on this model. He was Richard Trevithick, a huge, hairy brute of a man, six feet two inches tall, who had been a dunce as a schoolboy, but who now had two passions in life—wrestling and machinery. He was good at both. In 1796, at the age of twenty-five, he had completed a model locomotive of his own and experimented with it in his boarding-house room, presumably to the disrelish of his landlady. To see this diminutive machine Trevithick one day brought his cousin Vivian in the hope that he, better supplied with money, might pave the way for a larger machine. So matters turned out—what Matthew Boulton was to James Watt, Cousin Vivian became to Richard Trevithick. By the end of 1801 the two had completed a steam-driven machine. It had horizontal cylinders, the pistons of which were attached to cranks ninety degrees apart. It had a return tubular boiler and an automatic safety valve. It discharged exhaust steam up the stack—although Trevithick did not realize that this would increase the draft. For that purpose he had supplied bellows, driven by the crank axle. Pistons drove the wheels through crank axle and spur gear.

All this was good engineering; the result was that, on Christmas Eve, the Trevithick machine, with its queer excrescences of machinery, made its gasping way through the streets of Camborne (near the very tip of Cornwall), carrying several proud but self-conscious people at a speed of about five miles per hour. One of them left this testimony behind him:

> I knew Captain Dick Trevithick very well, I was a cooper by trade, and when Trevithick was making his steam carriage I used to go every day into John Tyack's shop at the Weith, close by here, where they put her together. In the year 1801, upon Christmas Eve, towards night, Trevithick got up steam, out on the high road, just outside of the shop. When we saw that Trevithick was going to turn on steam, we jumped up, as many as could, maybe seven or eight of us. 'Twas a stiffish hill going up to Camborne Beacon, but she went off like a little bird. When she had gone about a quarter of a mile there was a rough piece of road covered with loose stones. She didn't go quite so fast, and as it was a flood of rain, and we were much squeezed together, I jumped off. She was going faster than I could walk, and went up the hill about half a mile further, when they turned her and came back again to the shop.

Much encouraged, Trevithick and Vivian constructed a new machine, part locomotive, part stagecoach. It was built in Camborne but Trevithick wanted Londoners to see it. The two disciples of steam locomotion actually drove the new machine over the loose, rough roads to the distant city of Plymouth, to their own proud discomfort and to the terror of more than one toll-keeper along the route.

From Plymouth, Trevithick and Vivian took ship for London to practise the sweet arts of salesmanship upon citizens who might furnish them with further funds. The dismantled engine, which traveled with them, was reassembled again at Felton's Carriage Shop in Leather Lane, and Vivian ran it from there to Gray's Inn Lane, Lord's Cricket Ground, Paddington, and return—a distance of ten miles through the London streets. But Londoners remained quite certain that if a locomotive were worth while a Londoner would have built it; in the face of their chilly hostility, Trevithick sadly gave up his ambitious ideas and retired into South Wales.

But he could not give up his idea, particularly when South Wales abounded in coal mines with tramways which cried aloud for the services of his iron colt. He persuaded a mine owner to give his new machine a trial.

Here was the first locomotive to be put to a practical use. It was not so good an engine as his first; it had only one cylinder, and that of 8 inches diameter, with a prodigious stroke of 4 feet 6 inches. It had an impressive flywheel 9 feet 6 inches in diameter. With water and fuel it weighed five tons. Nevertheless it made a successful trip from Pen-Y-Darran to Plymouth on a coal tramway on February 10, 1804. Eleven days later it made its second trip, hauling behind it five "waggons" loaded with 70 passengers and 10 tons of bar iron. The passengers were once again treated to the breathless speed of 5 miles per hour and the train traveled 9 miles in 4 hours, including stops. Unhappily this brave great-grandfather of all railway locomotives was seized with a crotchet one day and to its great misfortune ran off its rudimentary track—burying itself sulkily in a ditch.

And into the ditch with it went Richard Trevithick. The Conservative Party had been waiting for just such an opportunity to show that new ideas lead inevitably to disaster. The clamor against Trevithick and his locomotive grew steadily louder until finally in 1816 this poor Cornishman fled in a mixture of bitterness and terror to the gold mines of Peru, which were stranger but seemed likely to be less unfriendly.

But even before his departure, Trevithick's virus had infected England. In 1811 John Blenkinsop of the Middleton Colliery near Leeds was granted a patent upon a new locomotive. Unfortunately he had decided that the friction of the drivers on the rails was not enough for traction and therefore added a rack laid at the left-hand side of the track with which a gear wheel on the engine engaged. This gear wheel was the driver, and it broke so frequently that it obscured most of the other virtues of Blenkinsop's locomotive. Missing the advantage of Trevithick's horizontal cylinders, Blenkinsop provided two 8-inch vertical ones set into the boiler to reduce heat losses, but he was more sensible than Trevithick concerning the stroke, which he held down to 20 inches. His machine, too, weighed five tons and could haul 94 tons at 3½ miles per hour. It was first put to work in August, 1812, and for close to twenty years thereafter it shuttled back and forth on the 3¾-mile tramway of the Middleton Colliery. It must have elicited blistering speeches in the House of Lords, but somehow it escaped destruction by any of the more zealous of the Lords spiritual and temporal who guarded the Realm from innovation. They were awaiting their great battle which would come later.

Meanwhile two other gentlemen began to bend their thoughts toward a better manner of carrying coals from Newcastle. They were employees of the Wylam Colliery in

Science Museum, London

THE "SANS PAREIL"

MODEL of Timothy Hackworth's entry for honors in the Rainhill Contest

Baltimore & Ohio Railroad Archives

"THE NOVELTY"

MODEL of the product of that inventive genius, John Ericsson, whose *Monitor* later helped the North to win the Civil War

Baltimore & Ohio Railroad Archives

"TOM THUMB" (MODEL)

PETER COOPER'S famous experiment, the first locomotive to be built in the United States

Smithsonian Institution

STEVENS'S LOCOMOTIVE (MODEL)

A GEARED design, developed in the United States by Colonel John Stevens

the most famous of all English mining districts, and their names were Timothy Hackworth, foreman of the smiths, and William Hedley, a mining engineer. To them the ramp-and-cog arrangement of John Blenkinsop seemed unnecessary. Either skeptical or unfamiliar with the previous achievements of Trevithick, they began work to prove what needed no further proof—that wheels on rails would provide sufficient friction for whatever pull a locomotive might develop. By loading a handcar with pig iron until it equaled the weight of their proposed steam machine, they proved their point in October, 1812. They set a crew of men to pump industriously at the levers and hitched coal cars on behind. Not until they had hitched on a great number did the wheels of their muscle-driven locomotive slip on the rail heads.

With that problem out of the way, they set about building a steam machine. For some four or five months they forged and hammered and fitted to the accompaniment of the amused and scornful jests of most of their associates. Early in the simple year of 1813 their handiwork stood complete. They had a locomotive whose name by a later christening, half derisive, half affectionate, was "Puffing Billy."

Unfortunately "Puffing Billy" in his original form was something of a weakling, and in many ways did not measure up to the engineering excellence of Richard Trevithick's first design. Yet he was stoutly constructed and had an honorable, if anemic, career. His cast-iron boiler was designed for a pressure of 50 pounds per square inch. He had one 6-inch cylinder and, like Trevithick's second engine, was equipped with a flywheel to smooth out the pulses of energy from the cylinder before they reached the driving wheels. This weak but willing machine could, when the wind was favoring, draw behind it at 5 miles per hour as many as 6 coal cars—each of about the size and capacity of one large farm tipcart. But the heating surface of the boiler proved too small for the ambition of the cylinder and this first "Puffing Billy" was soon retired for alterations.

But Hackworth and Hedley had confounded their enemies by producing a practical machine; straightway they set about with improvements. The result was a second "Puffing Billy" which appeared shortly after and which to-day, appropriately mounted, yet far from dead, stands in the Science Museum in South Kensington, London. This second "Puffing Billy" had not one but two vertical cylinders and could thus dispense with a flywheel. It had a wrought-iron boiler which supplied power for pistons, connected to the wheels not by a simple crank and connecting rod arrangement but by the walking-beam system which still distinguishes a certain elderly vintage of ferryboat. None the less, this second "Puffing Billy" operated for forty-seven years in drawing coal waggons from the Wylam Colliery to the river Tyne—a faithful iron colt that was not put out to pasture until the year 1862. Steam locomotion was established.

The locomotive was the child of the colliery, yet only the most progressive colliery, in the days of Hackworth and Hedley, would have anything to do with the idea of steam propulsion. Just so a century earlier had the collieries resented the first efforts of Thomas Savery and Thomas Newcomen to ease their horses of the endless treadmill duty of unwatering a mine.

When the eighteenth century expired, young George Stephenson was nineteen years old. He had been born on June 9, 1781. There were no child-labor laws in those days, and for as long as he could remember George Stephenson had earned his youthful living by driving a horse in the

West-Moor Colliery near Wylam. At the age of seventeen he had graduated to the position of plugman—for this mine was sufficiently progressive to have a pumping engine, and Stephenson's duty was to stand by the end of the engine's discharge pipe with a plug to stopper it when the water in the sump was all pumped out, so that the water remaining in the pipe would not run out to spoil the suction when the sump had filled and the engine was started again.

This George Stephenson was gifted with a mechanical ingenuity more than equal to the job of plugman. One of his earliest exhibits of ability came when one day he offered some entirely gratuitous advice to workmen at the High Pit Mine of the "Grand Allies" near the scene of his own labors. Its pump refused to work and young George diagnosed the trouble as in the engine. Everyone else knew better, but meanwhile water was continuing to rise and the mine was slowly being drowned. Finally on a Sunday morning the desperate engineer sent a call to George Stephenson to come and apply his treatment, since nothing else had sufficed. It then became apparent that, unsuspected by all, George was a young man who had been overcarefully reared. "Not to-day," he said in effect. "It is the Lord's Day. To-morrow, if you wish, I will repair your engine." He left the scene of his temptation with an immense dignity. The engineer, whose mine was filling, alternated between prayers and curses but had no recourse save to wait until the pious George chose to help him. Bright and early, with all the expedition of an Alger hero, George next morning repaired the engine and likewise succeeded in persuading unbelievers that the Lord's way, although occasionally dilatory, was usually the best.

George Stephenson's piety did not prevent him from picking up good ideas where he might. He, too, was now bitten by the idea of steam locomotion. Some years later he saw Richard Trevithick's second engine; to the inspiration of that was added his first glimpse of "Puffing Billy." He made sketches of the machine of Hackworth and Hedley, pumped an assistant with numerous questions, and by 1814 had built his first locomotive. Having convinced the owners of the same "Grand Allies" Colliery at Killingsworth that here was a money-saving adjunct for their business, he put his machine into operation on July 25th of that year.

The results were dreadful. Stephenson had given his first machine (called "The Blücher") a boiler 8 feet long and a little less than 3 feet in diameter, discharging through a 20-inch stack and supplying power to two vertical 8-inch cylinders connected to 3-foot drivers by means of cross heads and connecting rods to spur gearing. The task of this locomotive was to haul coal the nine-mile distance from Killingsworth to the nearest tidal basin over a three-foot-gauge track.

The failure of this first Stephenson locomotive is not expressed in the above statistics. Its great trouble was that after presenting itself as an economical substitute for horses, it then proceeded to cost more than horses do. It was a tired engine whose best performance was to haul 30 tons of coal up an incline of one foot to every 450. Since two good horses would find no trouble in pulling off this same trick without any show of bravado, "The Blücher" was not regarded as a success—perhaps because, among other defects, it had no springs, and the way of the transgressor, accordingly, was hard. "The Blücher" frequently complicated this by halting him in his path with a sudden lack of steam.

This combination of disabilities was enough for the owners of the "Grand Allies," who told George Stephenson that his idea would not do. Stephenson in his turn protested that all his locomotive needed was a new boiler, whereupon the mine's directorate snorted in traditional Conservative

Party fashion and told George Stephenson just what they thought he might do with his locomotive.

Instead of following this advice, Stephenson retired from coal-mining and set up a shop in Newcastle. Here he built a second engine which this time eliminated gearing and connected wheels to pistons by the simple device of a crank. Stephenson thus powered one set of wheels from the engine and then connected the axle with a second set by an endless chain running over sprockets. The Patent Office, on February 28, 1815, awarded him a patent upon this important part.

Stephenson, despite his uncertain successes, was now becoming a power in the new world of machinery. For one thing he was forceful and persuasive; for another he now had as an associate at the famous "Fourth Street Works" in Newcastle the same Timothy Hackworth who had collaborated on "Puffing Billy." Hackworth was a brilliant engineer, many of whose ideas are still attributed to Stephenson. By September 30, 1816, Stephenson had produced a third engine. It was a centipede of steam, with driving wheels coupled together by chains and sprockets—the wheels no longer simple castings but made of built-up wrought iron. And to make things even looser and noisier, the vertical cylinders were set on springs in the hope of reducing vibration produced by the heavy pounding of ill-balanced machinery upon uneven track. Later, in deference to Hackworth's sensibilities, these springs were removed and the six drivers reduced to four, this time coupled not by chains and sprockets but by side-rods.

Yet no Stephenson locomotive had by now been a real success. His boilers had always been inferior to those of Trevithick, and one early historian of steam declared over a hundred years ago that a Stephenson locomotive had seldom been capable of pulling a load up a track that inclined as little as eighteen feet to the mile.

But in 1818 the Hetton Colliery decided to use engines instead of horses to bring its products to the wharfside, and Stephenson, whose publicity was always good, was chosen to lay out the road, superintend the construction, and build the locomotives. He built five—a mammoth order—with improvements over his earlier Killingsworth design. On a level stretch, they would haul as much as 64 tons at 4 miles per hour. Stephenson and the colliery managers felt swollen with success, even though Stephenson found it necessary to resort to a cable system powered by stationary engines to haul cars up inclines by a rope and drum arrangement.

For all its defects, the eyes of England were now focused upon the locomotive. In the year 1821 a momentous bill passed Parliament, before the drowsing House of Lords realized what, in its somnolence, it had authorized. The bill permitted certain ambitious gentlemen to construct a property that was known as the Stockton & Darlington Tramway. Perhaps Parliament permitted this breach of the peace because the promoters first expected to use only horses as motive power and were doing little save providing a freight stage line with a track. George Stephenson, on the alert for new business, persuaded the directors that steam was the proper motive power; having won his point, he was retained to organize the work and furnish the machinery.

For four years Stephenson and his assistants toilingly cleared the way and laid rails over the twelve-mile distance between inland Darlington and Stockton near the estuary at Middlesborough. Stephenson, meanwhile, sweated over the iron entrails of a machine to be called "The Locomotion," which was to weigh 6½ tons, work at a pressure of 25 pounds to the square inch, carry 240 gallons of water, and haul a wooden tender filled with ¾ of a ton of coal. It would have four 48-inch driving wheels, coupled by side rods in Hackworth's fashion and connected to two vertical 10-inch cylinders, the pistons of which had a 2-inch stroke. In the fashion of the day these cylinders protruded from the top of the boiler in which they were immersed to reduce heat losses.

By now the suspicion was growing in the public mind that if an engine could haul coal it could also haul people, and although the Stockton & Darlington Tramway had at first intended to be nothing but a freight transporter, the directors, in answer to a public clamor, decided to venture further. The result was the first passenger railroad in history.

On the great day, September 27, 1825, the line was opened, and there was festival along the route. The train was ready to begin its tremendous journey with George Stephenson himself at the throttle of his engine. There was vast excitement throughout the country—for 90 tons of men and material were to set out on a voyage which might carry them as fast as 10 or 12 miles per hour. And here was the order of the first great parade of steam:

A horseman carrying a red flag, leading the cavalcade to warn of coming danger.

"The Locomotion," with George Stephenson in command, wearing a top hat.

Six waggons carrying coal and flour.

One sumptuous covered waggon loaded with the directors of the line. (The first six waggons which separated the directors from the engine were placed there to protect these dignitaries in case "The Locomotion" should explode.)

Twenty-one coal waggons carrying a total of 450 people—lesser lights than the directors, but still important enough to command tickets.

Six more waggons of coal.

Thronging the route of this parade was an excited crowd described in the language of the day as a "vast concourse."

The concourse was no more excited than George Stephenson, for the crucial moment of his engineering career had arrived. Could his engine really draw behind it almost fifteen times its own weight, or was he about to make a fool of himself, ruin his career and perhaps, in the gentle custom of the day, be tarred and feathered for his failure? Well, there was no escaping it now. A brass band in the rear coal waggons was blowing and pounding with all its might, the signal was given, the moment was at hand.

A prophetic horse wandered across the bows of this new caravel, perhaps in a faint desperate protest against the oncoming machine age. In savage nervousness Stephenson yanked his whistle lever and the horse, frightened out of a year's oats, vanished like the Boojum. The throttle was open; there was a moment of terrible protest in the internal organs of "The Locomotion," and—some 140 wheels began to turn. The great journey had begun. The brass band shrieked amid sparks and thunder, and for the next hour a cheering throng saluted the proud progress of the first passenger train—drawing behind it as many people as often travel at one time on a limited to-day. Without death or explosion or any sign from a displeased god, the train groaned to a stop in Darlington—and the stagecoach of history lay dying by the wayside.

Here at last was a clear-cut, unequivocal triumph; success for Stephenson and steam. On its initial run "The Locomotion" had sometimes reached a speed of fifteen miles per hour. Even after ten years of service it was able to run a ten-mile race with a horse-drawn mail coach and beat it by a few hundred yards. It served its masters until 1850, and to-day, honorably turned out to pasture, it stands upon a pedestal in the Bank Top Station in Darlington.

The success of "The Locomotion" presently made the Stockton & Darlington Tramway into a great transportation company. It had thirty-eight miles of single track, including three branches. So insistent was its clientele that the company was forced to run a daily passenger coach called, frankly enough, "The Experiment," which would hold six passengers inside and twenty outside. Two hours were necessary for a round trip.

George Stephenson had wisely made Timothy Hackworth general manager of the railroad when Hackworth had thought of leaving to set up a rival locomotive works. It was a doubly fortunate step. No sooner had the first railroad been constructed than one of the first vultures of economics settled down to batten on it. This was the vulture of Fixed Charges which still flaps its broad wings to-day. Steam engineering was a success in every way except financially. The managers of the Stockton & Darlington Tramway suddenly found themselves facing balance sheets which showed them that steam power was three times more expensive than horsepower. There were long faces, the railway stock fell to less than half its original value, and in the Conservative Party there was triumph. Nothing would supplant the horse.

Timothy Hackworth believed that much of this disappointment was due to improper engineering. For one thing "The Locomotion" used too much fuel. One day, at a melancholy meeting of the directors, Timothy Hackworth swallowed twice and said, "Gentlemen, if you will allow me to make you an engine in my own way, I will engage that it shall answer your purpose." Not too willingly, but because refusal might have had financial embarrassements, the directors bestowed an old boiler shell upon him and wished him godspeed. Hackworth retired to Newcastle and set to work on an engine in his own way.

The machine that he produced was amazing. It had the first decent boiler since the now forgotten Trevithick's. It had two 11-inch vertical cylinders which drove six 48-inch side-rod-coupled driving wheels. It had a boiler feed pump worked by eccentrics on the driving axle. It had self-lubricating bearings with oil reservoirs. It was the first properly spring-mounted locomotive, with the suggestion of present "equalizing beams." It used its exhaust steam as a jet beneath the fire grate to preheat the feed water and as a blast to increase draft up the stack.

This was the "Royal George." It was an immediate and spectacular success, which could haul 32 loaded cars weighing 130 tons at 5 miles per hour. In its first year it hauled 22,442 tons of goods over a 20-mile line at a total cost of $2,250. Horses, for this same work, would have cost $4,850. The transported directors rubbed their hands together and said, "All we want now is plenty of Timothy's locomotives." They were established.

Night and day for thirteen years the "Royal George" went on with its work; when it was eventually sold at second-hand, the price of $600 which it brought more than repaid its entire original cost.

Yet before the Stockton & Darlington Tramway had even begun its operations other companies were rising in emulation. There was the Monklands Railway in Scotland and the Canterbury and Whitstable in England. But none of these could compete in grandeur with a company which organized itself in 1824 with a backing of $2,000,000 and announced itself as the Liverpool & Manchester Railway. Once again George Stephenson was retained as engineer. This time, Parliament was not so easy. The Lords had awakened to the railroad as a threat to their revenues from canal companies. For more than two months Parliament agonized to determine whether or not a charter should be granted. "No," thundered the canal companies. "Such a venture roaring and clattering across the fields would ruin the countryside for fox hunting!" It was a powerful argument and even the adroit Stephenson scarcely knew how to meet it. Parliament pressed its advantage and heckled him for three days. Were not the red-hot chimney ends likely to frighten people? They might be so much red paint, said Stephenson. What would happen if a cow should stray upon the track? Would that not be an awkward circumstance? Awkward indeed, said Stephenson, for the cow.

Then—"You realize, of course," purred one suave member, "that no reputable engineer is willing to appear before Parliament to advocate the use of steam locomotives?" "*I* am," thundered Stephenson. His rage at the insult betrayed him. When another member demanded to know what speed his craft would make, he unguardedly told the truth: twelve miles per hour was quite possible.

Here was a bad tactical blunder. A speed like this was too fast for the Lords, who felt that eight miles per hour was the uppermost limit. They drew back in chill disapproval and vetoed the railway bill. England must be kept safe for travelers and serene for foxes.

Undismayed, the promoters returned again. A new bill was drafted and introduced. Lobbyists poured out more than $100,000 to see it through. The Lords were spent and drowsing now; the rash Mr. Stephenson was kept out of sight, and the new bill, which slyly made no specification of the motive power to be used, was passed on May 1, 1826. There would be a railroad.

CHAPTER XII

STACK BLAST

GEORGE STEPHENSON went to work again. There were thirty-one miles of track to be laid, eighteen of them upon stone blocks instead of wooden sleepers. There were four million tons of stone and soil to move to drive the railway through. There were some millions of dollars to be spent.

To supply motive power for this railway, the directors felt disinclined to trust any one man. They published the specifications of what they wanted, and announced a contest of locomotives from which the winner would be picked. So began the great Rainhill Contest. These were the specifications:

> Railway Office, Liverpool, 25, April 1829.
>
> Stipulations and Conditions on which the Directors of the Liverpool and Manchester Railway offer a premium of £500 for the most improved Locomotive Engine:
>
> 1st. The said engine must effectually consume its own smoke, according to the provisions of the Railway Act, 7, Geo. IV.
>
> 2nd. The engine, if it weighs six tons, must be capable of drawing after it, day by day, on a well-constructed Rail-way, on a level plane, a train of carriages of the gross weight of twenty tons, including the tender and water-tank, at a rate of ten miles per hour, with a pressure of steam on the boiler not exceeding 50 lbs per square inch.
>
> 3rd. There must be two safety valves, one of which must be completely out of the control of the engine-man, and neither of which must be fastened down while the engine is working.
>
> 4th. The engine and boiler must be supported on springs, and rest on six wheels, and the height from the ground to the top of the chimney, must not exceed fifteen feet.

5th. The weight of the machine, *with its complement of water* in the boiler, must, at most, not exceed six tons; and a machine of less weight will be preferred if it draw after it a proportionate weight; and, if the weight of the engine, &c., do not exceed five tons, then the gross weight to be drawn need not exceed fifteen tons, and in that proportion for machines of still smaller weight; provided that the engine, &c., shall still be on six wheels, unless the weight (as above) be reduced to four tons and a half, or under, in which case the boiler, &c., may be placed on four wheels. And the Company shall be at liberty to put the boiler fire-tube, cylinders, &c., to a test of pressure of water not exceeding 150 lbs per square inch, without being answerable for any damage the machine may receive in consequence.

6th. There must be a mercurial gauge affixed to the machine with index-red shewing the steam pressure above 45 lbs per square inch.

7th. The engine to be delivered complete for trial at the Liverpool end of the Rail-way not later than the 1st October next.

8th. The price of the engine, which may be accepted, not to exceed £550, delivered on the Rail-way; and any engine not approved to be taken back by owner.

N.B. The Rail-way Company will provide the engine tender with a supply of water and fuel, for the experiment. The distance within the rails is four feet eight inches and a half.

No such steam-driven paragon was in existence, and between April and October, 1829, there were ten attempts to construct one. Most of these fell by the wayside. An inventor named Brandreth tried to compete by entering an affair called "The Cycloped"—a three-ton complication of wheels, platform, gears, belting, and treadmill. Upon the treadmill two horses were to gallop proudly but fixedly, spurred by a mirage of carrots. So it was that when October, 1829, arrived there were submitted in the Rainhill Contest but four machines—the "Rocket" of George Stephenson; the "Novelty" of Braithwaite and Ericsson; the "Sans Pareil" of

Timothy Hackworth; and the "Perseverance" of Burstall. The name of this latter engineer was an ominous one in a day of weak boilers. Ill fortune swiftly attended Burstall's entrant, which was damaged during unloading at Rainhill and shortly afterward withdrawn, cranking slowly away, in the words of one succinct eye-witness, "like an old wicker-work pair of panniers on a cantering cuddy ass."

The three nervous locomotives that remained were gay sights. The "Novelty" of Braithwaite and Ericsson was, in one description, "covered with copper like a new tea urn, which tended to give her a parlor-like appearance." Her woodwork was blue. The "Rocket" of George Stephenson and his son Robert was painted a sulphurous yellow and a pitch black, the whole surmounted by a white smokestack. The "Sans Pareil" of Timothy Hackworth was green, yellow, and black.

The vested interests that controlled canals and turnpikes looked sourly upon the Rainhill Contest. The canal which bore the commerce between Liverpool and Manchester was a poor means of connecting so great a port with so great a manufacturing center. It froze in winter; in summer the water fell so low that the barges could carry but half a load. If some clever engine builder were able to meet the conditions of the contest the canal might find itself with a formidable rival. In a few privileged chests there were a few apprehensive shudders.

The public at large took another view and flocked to the contest as if it were the Derby—which, in many respects, it was. Some ten to fifteen thousand people turned out, and along the line of test track there were grandstands which cost five shillings to enter.

The heavens did not smile upon the Rainhill Contest when it began on October 6, 1829. The three locomotives were weighed in like prizefighters before a battle; they were to be required to pull a load three times their own weight. But scarcely were these preliminaries over, and the "Rocket" and the "Novelty" brought out for a few preliminary canters, before rain fell generously upon the just and the unjust and the contest was postponed until the next day. The crowd, which had seen the "Novelty" dash down its track at a full 28 miles per hour, retired to the tavern at Kenrick's Cross, to wash down its amazement with a little ale.

The following day was just as inconclusive. The "Novelty," in a preliminary work-out, drew a load thrice its own weight at 20½ miles per hour, but one disappointment or another prevented much else from happening. The judges decreed that further trials be postponed until the next day, and the crowd, muttering faintly, went home.

Upon the next day the "Rocket" submitted to a number of indignities in the precise manner of the conditions fixed by the judges. The test track was a mile and three quarters long. The judges had specified that engines must be capable of performing ten round trips at an average speed of 10 miles per hour—ten round trips approximating the distance of 35 miles that separated Liverpool from Manchester. The "Rocket" did this stint in 3 hours and 10 minutes, thus averaging a speed of 11 miles per hour. Later in the day, with fresh fuel, the "Rocket" bettered its record with an average of 12 miles per hour and occasional bursts of 18 or 20 miles an hour maximum speed.

Then the "Novelty," with Braithwaite at the throttle, started out gayly, her copper sides gleaming prettily. On the third lap of the speed demonstration she was making 16 miles per hour when a feed pipe burst. Simultaneously, the heavens did the same, and the Rainhill Contest was rained out once more.

On the following day there were no races: the judges generously gave the time to Braithwaite and Ericsson who needed it for repairs and adjustments to the "Novelty." Next day the "Novelty" raised steam in 54 minutes—a fast time—and once again started off proudly, only to have her boiler wither with the heat of its fire and leave her dead. Whereupon the "Rocket," minus train and tender, whizzed along the track at 35 miles per hour and the public, delighted but a little frightened, went home to await the sixth day of the contest upon the following Monday, October 13th.

It was then the "Sans Pareil" made an appearance and, although overweight, was permitted to demonstrate her capabilities. They were considerable: she pulled a load of 25 tons behind her at 12½ miles per hour average, and frequently made 15, but after two hours disability seized her also when a feed pipe burst. She was a hard rider; her pistons were vertical and she pounded badly. The pistons of Stephenson's "Rocket" were inclined at forty-five degrees. It was only John Ericsson, the pistons of whose "Novelty" were horizontal, who had the proper idea.

Unfortunately, on the following day, the "Novelty" once again blew her boiler, and Ericsson and Braithwaite retired from the competition. The "Sans Pareil" had likewise been through vicissitudes; a cylinder jacket would give way or the blower bellows would expire. So it was that by October 14th Stephenson's "Rocket" remained the only iron horse sound of wind; through all the troubles of the week she had sailed serenely, performing adequately under the most trying circumstances.

The "Rocket's" great advantage lay in her tubular boiler. It had been suggested to Stephenson by an official of the Liverpool & Manchester Railway—and had been patented not more than a year before, unknown to the English, by the French engineer Seguin. It enabled her to evaporate much more water with a reduced fuel consumption and gave

Smithsonian Institution

THE "BEST FRIEND" GOES RAILROADING

An old print showing the services of the first locomotive in the United States built for regular passenger service on a railroad

Smithsonian Institution

THE "WEST POINT"

A mixture of triumph and caution; note the barrier car behind the locomotive's tender—placed there for protection against boiler explosions

Smithsonian Institution

"OLD IRONSIDES"

Model of the businesslike locomotive which changed Matthias Baldwin's profession from watch-making to engine-building

Baltimore & Ohio Railroad Archives

THE "YORK"

A model made after alterations of the design which in 1831 won a prize from the Baltimore & Ohio Railroad

John Dixon (Stephenson's resident engineer) the opportunity of derogating Hackworth's "Sans Pareil":

> She burns nearly double the quantity of coke that the "Rocket" does and rumbles and roars and rolls about like an Empty Beer Butt on a rough pavement. . . . She is very ugly and the boiler runs out very much.

But Dixon did have the grace and fairness to add:

> Timothy openly accused all George Stephenson's people of conspiring to hinder him, of which I do not believe them innocent.

Perhaps George Stephenson's people did not actually hinder Timothy Hackworth, their former partner, but they certainly were not averse to taking a few pointers from him. Hackworth had hit upon the idea of blowing the steam blast up the stack, not, like Trevithick, because he could not think what else to do with it, but for the positive purpose of enhancing the draft. Stephenson knew nothing of this innovation until he began to wonder why Hackworth's sparks blew so violently from the chimney. Twenty-four hours after that, the "Rocket" appeared with a similar exhaust blast—it was even rumored that two of Stephenson's men learned the secret when they broke into the shed in which the "Sans Pareil" slept at night.

Notwithstanding, the "Rocket" showed itself the best record-making machine and the best combination of engineering advantages. Although no one knew it, a successful locomotive depended upon four things: a tubular boiler, the stack blast, cylinders that were horizontal and which had a direct connection between themselves and drivers. Only Stephenson's "Rocket" possessed enough advantages. Thus it was that after the "Rocket" had made 40 trips carrying 17 tons at a maximum speed of over 24 miles per hour and an average speed of almost 13, and had otherwise demonstrated its consistent abilities, the judges purchased it as the means of breathing the breath of life into the Liverpool & Manchester Railway. They bestowed upon Stephenson the prize of £500, which he shared with that same Mr. Booth who had proposed the tubular boiler idea. The "Rocket" was obviously foreordained for success; it had hauled 40 passengers at 20 miles per hour, whereas 12 miles per hour, as a promise from its designer, had been enough to frighten the House of Lords out of a year's sitting. And when rebuilt, ten years later, it further confounded the House of Lords by running 60 miles per hour in the East Cumberland Contest.

The happy Stephenson received orders for the construction of a number of duplicate locomotives. The first round trip over the line from Liverpool to Manchester was made on June 14, 1830, in a train drawn by the "Arrow," a Rocket-type engine. It traversed the 31 miles in 3½ hours, hauling a train consisting of two passenger carriages and seven freight waggons to the total sum of 39 tons.

It was not until September 15, 1830, that the railway was formally opened to traffic. Once again George Stephenson stood in his magnificence at the throttle of the "Northumbrian," a locomotive with the hitherto unsurpassed strength of 9 horsepower. This day, like the modest Stockton & Darlington day of a few years before, was marked by brass bands, flag-waving, and enthusiasm—in the quantity of these virtues all records were surpassed. The grandeur of the occasion was made almost insupportable by the presence of the Iron Duke, Wellington, who had met the great Napoleon at Waterloo and had, with some help, defeated him. The Liverpool & Manchester Railway was the first of any to be double-tracked; on this opening day traffic on both tracks flowed from Liverpool to Manchester. Upon Track 1 there traveled the "Northumbrian," carrying the band and literally tons of celebrities, including Wellington. On Track 2 traveled eight other trains, each powered with a new engineering triumph built by the Stephensons since the Rainhill Contest.

The day had its tragedy; Huskisson, whose influence had been greatest in persuading Parliament to give the railroad a charter, was struck by the oncoming "Rocket" at Parkside, and although the "Northumbrian" was uncoupled to speed him to the Manchester hospital at thirty-six miles an hour, he died, the first martyr to what was to be a growing mania in the world—the mania for speed. But the day of his death was also a day of great history—for the steam-powered railroad was established in England, and was soon to be a commonplace.

In its mysterious way the current of invention had, even before this, reached the youthful United States of America, where the railroad had first seen the light in the year 1828. Thomas Jefferson had assured everyone in 1803 that it would be one thousand years before the region east of the Mississippi would be fully settled. Had people been compelled to rely only upon turnpikes and natural waterways for their avenues of commerce it is quite probable that Jefferson would have turned out to be right. The power of steam had not occurred to him.

It had occurred to others. Construction of the first large-scale railway—the Baltimore & Ohio—began on July 4, 1828. It was to be some twenty-five years later that the first train on this railroad was to reach its final objective—the Ohio River, three hundred miles away at Wheeling, West Virginia. Baltimore on this century-old day was a festive city none the less. Charles Carroll, the last living signer of the Declaration of Independence, thrust a quavering shovel to move the first earth for the new construction. The city resounded with cheers and the firing of cannon.

It took some two years to push the Baltimore & Ohio's line as far as Ellicott's Mills, but on May 24, 1830, this thirteen-mile stretch was ready for operation. "Now," said Peter Cooper, "we need a locomotive." He was a director of the road, and in a position to make demands. "Oh, no, we don't," said the other directors. "We have a fine roadbed, and horses can very economically draw the cars over the rails, supplying all that can be wanted in the way of celerity with a minimum of expense."

Peter Cooper didn't believe it and was undismayed at finding himself a heretic. His fellow directors had sought to dispose of the locomotive by arguing about it. Peter Cooper, believing that valve actions would speak louder than words, began to build a curious engine, the size of a modern hand-car, which he called the "Tom Thumb." He made it with a vertical boiler, with tubes that were sawed-off sections of old musket barrels. At a pressure of 90 pounds to the square inch, these would, he hoped, deliver 1½ horsepower to a single cylinder 3½ inches in diameter by 14½ inches long. He supplied the "Tom Thumb" with 52 gallons of water and 1,190 pounds of coal and now found that his 1½ iron horses weighed the disquieting amount of 10,800 pounds.

Peter Cooper was doing a brave thing. His "Tom Thumb" needed not only to surmount hostile public opinion: it needed to negotiate an extraordinarily difficult roadbed. The early Baltimore & Ohio line had some fearfully sharp curves —detours around occasional rocks of moderate size that sprinkled the terrain between Baltimore and Ellicott's Mills. But Cooper was fired not only by the faith of engineering but by the zeal, which moveth mountains, of the realtor. He owned large quantities of land near Baltimore and shrewdly believed that the development of his property depended upon the success of the Baltimore & Ohio Railroad.

The trial of the "Tom Thumb" came on August 28, 1830. It exceeded even its builder's hopes by drawing 36 people in a passenger car (and 4 more on itself) at a maximum speed of 18 miles per hour. It made the trip from Baltimore to Ellicott's Mills in one hour and 12 minutes. The excited gentlemen of the party pulled out their notebooks and wrote their names in shaky handwriting to prove to skeptics that even at that thunderous speed it was still possible not only to breathe (there had been some doubts about this) but also to write. The return trip was an even greater triumph. The "Tom Thumb" traversed the distance in 57 minutes. Steam power for transportation in the United States was established.

But not too firmly established. The firm of Stockton & Stokes, stagecoach proprietors, was not prepared to relinquish their glory without a struggle. The pride of their stables was a gray horse, frequently used to pull the B. & O. coaches on the right-of-way. Forthwith, as Cooper's engine was returning on the day of its trial trip, Stockton & Stokes challenged the "Tom Thumb" to a duel. Peter Cooper accepted. Shortly the "Tom Thumb" and the gray horse stood ready for a race on parallel tracks to Baltimore from the Relay House. What happened is vividly preserved in a contemporary account written by that eminent engineer, John H. B. Latrobe:

> At first the horse had the best of it; for his steam would be applied to the greatest advantage on the instant, while the engine had to wait until the rotation of the wheels set the blower to work. The horse was perhaps a quarter of a mile ahead, when the safety-valve of the engine lifted, and the thin blue vapor issuing from it showed an excess of steam. The blower whistled, the steam blew off in vapory clouds, the pace

increased, the passengers shouted, the engine gained on the horse, soon it lapped him; the silk was plied, the race was neck and neck, nose and nose. Then the engine passed the horse, and a great hurrah hailed the victory. But it was not repeated; for just at this time, when the gray's master was about giving up, the band which drove the pulley, which moved the blower, slipped from the drum, the safety-valve ceased to scream, and the engine, for want of breath, began to wheeze and pant. In vain Mr. Cooper, who was his own engineer and fireman, lacerated his hands in attempting to replace the band upon the wheel; in vain he tried to urge the fire with light wood. The horse gained on the machine and passed it; and although the band was presently replaced, and steam again did its best, the horse was too far ahead to be overtaken, and came in the winner of the race.

Despite Peter Cooper's chagrin, his ghost now has the great satisfaction of knowing that, in the flesh, he won that race, after all. The Baltimore & Ohio Railroad was now in operation as the first regular passenger road in the United States and the first to operate with an American-built locomotive.

The citizens of South Carolina were not to be outdone by the citizens of Baltimore. Even in 1827 they had obtained a charter for a railway which they proposed to build to connect Charleston with Hamburg, 136 miles away on the western border of South Carolina. Having completed a portion of their roadway, they first offered a prize of $500 for the best locomotive which would use the power of horses on a treadmill. This was won by C. E. Detmold. Later they tried sails as a source of power. Both these attempts are described elsewhere.

Denied much comfort from these natural sources, they turned in desperation to steam. Deciding that they needed more than local talent to solve their problem, they called upon Horatio Allen. He had had experience with some locomotives imported from England to haul coal from mine mouth to navigable water between Carbondale and Honesdale, Pennsylvania, in 1828, for the Delaware & Hudson Coal Company. Allen studied the South Carolina project for thirty days and on January 14, 1830, submitted a magnificent report. Allen compared the cost of horses with the cost of steam and frankly admitted that horses showed an advantage. But he pointed out this, which had usually escaped notice: it was not likely that the horse would improve his mechanical efficiency; the energy which he supplied by burning hay and oats was not susceptible of man-made improvement, whereas the steam engine was still young and the ingenuity of man might quite reasonably, some day, be expected to gain an advantage over the digestive apparatus of the horse.

This cogent argument won the day. The Board of Directors of the South Carolina Railway unanimously voted that their railway should henceforth be developed *exclusively* for steam transportation. No such radical stand had ever yet been taken by any road. Then, exhausted with good works, the board thanked Allen for his report and turned over the task of building a locomotive to E. L. Miller, one of the directors. Miller, in turn, hired Mr. Detmold, the designer of the previously unsuccessful horse locomotive, who now went to work at the West Point Foundry in New York City with the help of David Matthew, the foundry superintendent. So it was that on October 23, 1830, barely nine months after Allen's report, the ship *Niagara* made port at Charleston with the locomotive, "Best Friend of Charleston," in its hold.

The "Best Friend" had four wheels with iron hubs and iron tires held together by wooden spokes, and with an iron web for the connecting-rod pins. The wheels thus coupled together were 4 feet 6 inches in diameter. They were driven by inclined cylinders, 6 inches in diameter and of 16 inches stroke. To drinkers of malt liquor the vertical boiler resembled a "porter bottle." The "Best Friend" weighed 4½ tons, had a boiler pressure of 50 pounds to the square inch, could exert a tractive force of 400 pounds, and at 20 miles per hour developed a full 12 horsepower. Her maw consumed vast quantities of pitch pine wood for fuel.

On January 15, 1830, only one day after Horatio Allen had submitted his report, the directors had begun building their roadbed. By the time the "Best Friend" was ready for service, seven miles of track were finished, and the locomotive was set to work hauling materials for further construction. In this prosaic service it moved mountains of wood and iron, and by the time the road celebrated its first anniversary in 1831, the United States Army saluted it with cannon and band music as it rolled along the track. Its friendship with Charleston might have continued unbroken had it not been for the overt act of a colored gentleman whose deplorable action and its consequences were sadly set forth by the Charleston *Courier* of June 18, 1831:

> The locomotive, Best Friend, started yesterday morning to meet the lumber cars at the forks of the road, and while turning on the revolving platform the steam was suffered to accumulate by the negligence of the fireman, a negro, who, pressing on the safety-valve, prevented the surplus steam from escaping, by which means the boiler burst at the bottom, was forced inward, and injured Mr. Darrell, the engineer, and two negroes. The one had his thigh broken, and the other received a severe cut in the face and a slight one in the flesh part of the breast. Mr. Darrell was scalded from the shoulder-blade down his back. The boiler was thrown to the distance of twenty-five feet. None of the persons were dangerously injured except the negro, who had his thigh broken. The accident occurred in consequence of the negro holding down the safety-valve, while Mr. Darrell, the engineer, was assisting to arrange the lumber cars, and thereby not permitting the necessary escape of steam above the pressure the engine was allowed to carry.

The casualties to the "Best Friend" were likewise severe. The mangled remnants were carried off to the shop of Thomas Dotterer and from them a revised locomotive emerged. She had outside cylinders and could thus be equipped with a straight axle instead of one with a crank in the middle. Moreover, the wooden spoke wheels were replaced by wheels made entirely of cast iron. The result was a great improvement. There was some philosophic doubt as to whether the engine was still the "Best Friend of Charleston" or another one; a dilemma escaped by a rechristening which named her the "Native." But the name did not stick; popular fancy demanded that she be known, having risen again from her own ashes, as the "Phœnix."

Meanwhile a new locomotive had arrived from the West Point Foundry, this time with a horizontal boiler as suggested by Allen, and 6 fire tubes, 2½ inches in diameter and 6 feet long. The management was now taking no chances. When this new locomotive, called the "West Point," went into service, she carried a new type of safety valve, out of reach of anyone save the engineer. Behind her on her trial trip she hauled a "barrier car," consisting of six bales of cotton strapped with iron hoops. This, said officials, would be a feature of all trains for protection of passengers *when* (there were no *ifs* about it) the boiler exploded. Behind the barrier car on the trial trip came the brass band whose departure from modern railroading has taken so much of the color out of daily life. Then came 117 passengers, including 50 brave ladies and 9 foolhardy technical gentlemen who rode upon the engine, anxiously watching the boiler for symptoms of an explosion. But the "West Point" never exploded. She wore out in service, and after a while the barrier car vanished and left railroading to develop unhindered.

These Southern developments had not escaped the North.

Albany and Schenectady, N. Y., now decided that they needed connective tissue. The Mohawk & Hudson Railroad was the result. This famous road, the great-grandfather of the present-day New York Central Railroad, was formally opened on August 9, 1831, when the "De Witt Clinton," likewise built at the West Point Foundry by David Matthew, made its first round trip over the 12½ miles of completed track that now ran out of Albany. The "De Witt Clinton" had 2 cylinders, 5½ inches in diameter and 16 inches in stroke. Inside cranks were attached to the forward pair of the four wheels, each 4 feet 6 inches in diameter. The boiler was a large dome, with 30 copper tubes designed to operate at a pressure of 50 pounds to the square inch. The "De Witt Clinton" had originally been designed, alone among the engines of the day, to burn anthracite, but her draft proved inadequate and so pitch pine served once again as fuel. Her total weight was 3½ tons without water.

Behind her, on this famous day, were five coaches, each resembling a stagecoach running on iron wheels. David Matthew was at the throttle, and John T. Clarke, as "Master of Transportation," then and there founded the honorable line of railway conductors. Matthew stoked the firebox of the "De Witt Clinton" mightily, and while steam gathered in her boiler, ticket holders swarmed into and upon the coaches. Upon the "platform car" behind the "De Witt Clinton" stood a wooden cask of water and a supply of pitch pine wood. When at last the locomotive manifested a willingness to start, John Clarke mounted to the platform car, placed a horn to his lips and split the heavens with a piercing blast.

The nervous Matthew yanked his throttle open; the loose couplings, one by one, took up their slack, and in groups of fifteen and twenty, the illustrious necks of New York's political life snapped back hard. Then, as the scrimmagers untangled themselves, the "De Witt Clinton" began pouring back upon the roofs of the carriages a heavy mixture of smoke and cinders. The passengers had equipped themselves for such inconveniences with umbrellas. They raised them now, but the showering sparks of the engine set first one and then another ablaze. The train swept on to its destination amid a grand carnival of fiery parasols. Train and passengers presented, in the words of one Judge Gillis, who was aboard, "a very motley appearance on arriving at the first station."

Moreover, the excitement by the roadside was fully as great as on the train. As always with these new spectacles, the countryside had turned out to witness. Unaware of the hazards, farmers parked their buggies close to the right-of-way, so that the "De Witt Clinton" progressed continuously through a churning mob of frenzied horses, teetering carriages, and the symptoms of panic. The trip ended in a general welter of blackened faces and shattered nerves—for the speed of the "De Witt Clinton" had sometimes exceeded 25 miles per hour. But once again, the railroad had driven a spike into the coffin of the horse.

It was a great year for the railroads. In January, 1831, the Baltimore & Ohio offered a $4,000 prize for the best locomotive to be built in the United States, capable of drawing 15 tons at 15 miles per hour. There were five entries; although none of them was considered entirely satisfactory, the prize was eventually awarded to the famous "York," built by Davis & Gartner of York, Pennsylvania. She had a vertical boiler and, in spite of the lessons of John Ericsson, vertical cylinders, which made her pound horribly. But she burned anthracite without indigestion, and could negotiate a curve of four hundred feet radius. It was considered remarkable.

The Baltimore & Ohio contest had a stirring effect upon the practical men of the day. There was one Matthias Baldwin—no engineer at all, but a jeweler. In 1831 he had made a miniature locomotive for the Philadelphia Museum which would actually work. Then, in 1832, the yearning rose in him to build a full-scale engine for the Philadelphia, Germantown & Norristown Railroad, which had none and depended upon horses. He had no tools; the famous Baldwin Locomotive Works began when Matthias Baldwin grimly set about boring his locomotive cylinders by fixing a chisel in a block of wood and turning it by hand.

The eventual result was "Old Ironsides." The directors of the railroad at first nearly rejected her, and the heavy-hearted Baldwin vowed then that he would stick to jewelry henceforth. The boiler was too small; on her trial trip passengers had alternately to push and ride. She was too heavy. She reversed with the greatest difficulty. Her water pump would not work. Yet one by one her lunacies were corrected, and shortly the Philadelphia, Germantown & Norristown Railroad could advertise thus:

> NOTICE TO THE PUBLIC
>
> The engine with a train of cars, will be run daily, commencing this day, when the weather is fair. When the weather is not fair, the horses will draw the cars. Passengers are requested to be punctual at the hour of starting.

"Old Ironsides" was the one locomotive which the road possessed. But everywhere, now, practical men were turning their minds and hands to the problem of motive power. Within a year locomotives were sprouting luxuriantly from the seed of "Tom Thumb," and it would be only a short time now before the age of empire-building would begin.

CHAPTER XIII

John Fitch Writes in Water

SPAIN would like to have it thought that the harbor of Barcelona in 1543, a bare half-century after Christopher Columbus had sailed from Genoa, saw the power of steam first applied to sea craft. In that year Blasco De Garay is supposed to have attached paddle wheels to a vessel of "two hundred tons burden" in one part of which, and subtly connected with the paddle wheels, was a "vessel of boiling water."

But Dionysius Papin—counselor and physician to his royal highness, elector of Cassel, and professor of mathematics at the University of Marburg—had a better claim. Papin had built a simple piston engine, elsewhere described, and in 1690 he proposed to connect it to paddle wheels. By 1707 he had actually equipped a boat with his proposed machinery—the piston engine pumped water against the vanes of a water wheel; these in turn drove the paddles of the ship. So, at least, it went in theory, and Papin's vessel rocked quietly on the river Fulda at Cassel, while this early marine engineer sought to enlist the influence of his friend Leibnitz, the mathematician, in having the electoral councilors of Hanover grant him permission to sail his craft down the river Weser to Bremen. Hitherto, ships from Cassel, or any point on the Fulda, were not permitted to enter the Weser, but must instead unload at Münden. Papin wished to navigate his ship unmolested through the Electoral Domain.

The electoral councilors felt otherwise. When Leibnitz broke the seal upon their reply he read the equivalent of this:

> The Electoral Councillors have found serious obstacles in the way of granting the above petition, and, without giving their reasons, have directed me to inform you of their decision, and that, in consequence, the request is not granted by His Electoral Highness.
>
> H. REICHE.
>
> *Hanover, July 25, 1707.*

Papin gave up seagoing forthwith. The chilliness of His Electoral Highness had been matched by the warmth of a mob of boatmen on the Fulda, who saw the ruin of their business in this embryo steamship and "utterly destroyed it." So Papin went to England, where the seas were freer.

In 1736 one Jonathan Hulls took up the fight and obtained an English patent for the use of a steam engine in a ship. One year later he published a pamphlet which made it evident that his invention consisted of a Newcomen engine, fitted with a counter-weight, a system of ropes and grooved wheels, and a ratchet-action to convert the reciprocating motion of the Newcomen piston into the rotary motion of a paddle wheel. He wrote:

> In some convenient part of the Tow-boat there is placed a Vessel about two-3rds full of water, with the Top closed; and this Vessel being kept Boiling, rarifies the Water into a Steam, this Steam being convey'd thro' a large pipe into a cylindrical Vessel, and there condensed, makes a Vacuum, which causes the weight of the atmosphere to press down on this Vessel, and so presses down a Piston that is fitted into this Cylindrical Vessel, in the same manner as in Mr. Newcomen's Engine, with which he raises Water by Fire.

He proposed timber guards to protect his paddle wheels from injury and thought further to attach crank legs to his paddle shafts so that in shallow water the legs would strike the bottom of the river and propel the ship forward with even greater force. And so he wound up:

> . . . Thus I have endeavored to give a clear and satisfactory Account of My New-invented Machine, for carrying Vessels out of and into any Port, Harbor, or River, against Wind and Tide, or in a Calm; and I doubt not but . . . that the Scheme I now offer is Practicable, and if encouraged will be useful.

Mr. Hulls was not encouraged and his scheme was accordingly not useful.

But the idea persisted. In 1752 the French Academy of Sciences offered a prize for the best essay on a means of propelling vessels without wind. This was won by that famous scientist Daniel Bernouilli, who proposed either a reaction jet of water, or oars at the stern and sides of a vessel, driven by steam or animals on a treadmill.

Then there appeared a paper by the Abbé Gauthier, published at about the same time by the Société Royale des Sciences et Lettres in Nancy. The Abbé argued mightily for the steam engine as opposed to all other sources of power. A galley propelled by 52 oars, he said, travels 5 miles per hour and, allowing 5 shifts, requires a crew of 260 men. A steam engine of equivalent power would weigh less, take less space, be less expensive and require fewer repairs. He proposed iron bands around the boiler of his steam engine to prevent an explosion and thought, in a day innocent of marine engineering, that sea water might efficiently be used to raise steam. He would like to see the walking beam of a Newcomen engine connected by its chain to a pawl-and-ratchet arrangement which would drive the paddles. But despite his fervor, this suggestion too was buried under mountains of indifference.

Then in 1760, a Swiss, A. J. Genevois, published in London a treatise called *Some New Enquiries Tending to the Im-*

provement of Navigation. His idea was that steam power should be used to compress heavy springs, which when placed in the ship would drive the paddle wheels.

In the same year of this suggestion England received a significant visit from one William Henry, a citizen of Lancaster, Pennsylvania, in the Colonies, the inventor of two new boons, the rag carpet and the screw augur. He had been born in 1729 and was related to that gentleman of the same family name who, some years later, was to demand the option of liberty or death. William Henry had visited England in 1760 on business and there had been much impressed by the growing power of steam. He saw in it an adjunct to navigation. On his return to America he built an engine in 1763, installed it in a boat fitted with paddle wheels, and launched it on the Conestoga River. Unfortunately the weight of the boat was too much for its displacement, and in situations like that a boat can do only one thing. William Henry's boat did it. It sank.

Henry recovered from his disappointment sufficiently to present a steamboat design to the Pennsylvania Philosophical Society some time later. His services to steam navigation, however, were less important for the things he accomplished than for the visitors he had. Among his friends were two young men whose names were John Fitch and Robert Fulton. The line of succession had been established.

Naturally unaware of proceedings in Lancaster, Pennsylvania, the Comte D'Auxiron formed in France a company which began in 1770 to raise money to build two steamships. Affairs progressed with great enthusiasm, and Louis XV promised to Comte D'Auxiron a monopoly on steam navigation for fifteen years should his efforts be successful.

By September, 1774, one steamship had been launched and was being fitted out—only to spring a leak and founder at her wharf. The noble lords with whom the count had been

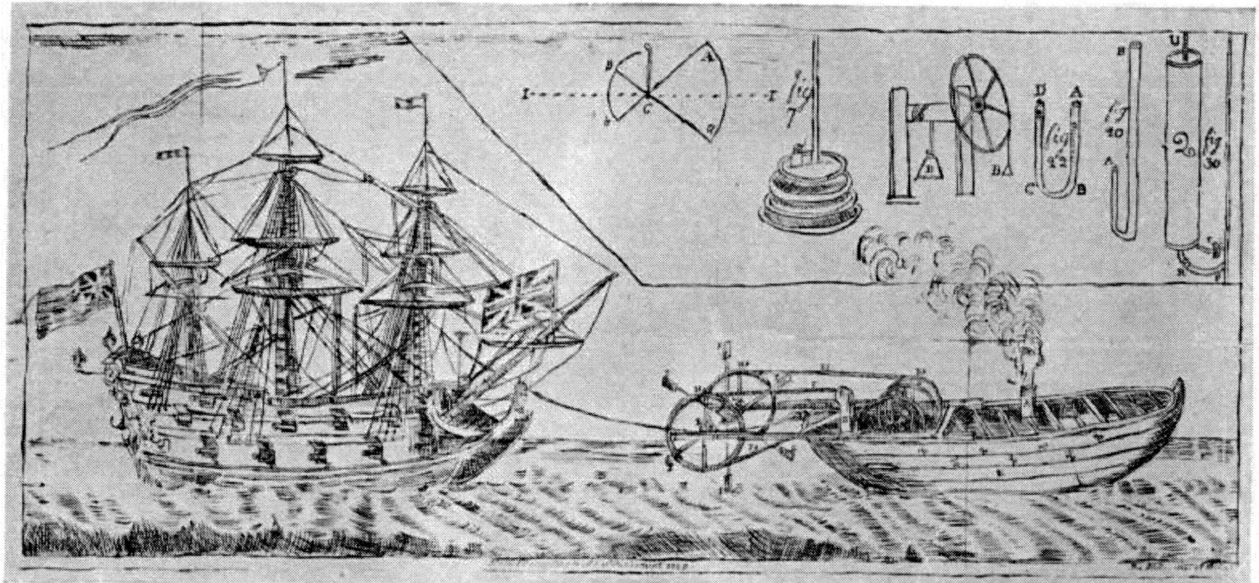

Smithsonian Institution

BRAVE JONATHAN HULLS

A SHEET from his draft published by Act of Parliament in 1737, proposing navigation by means of steam

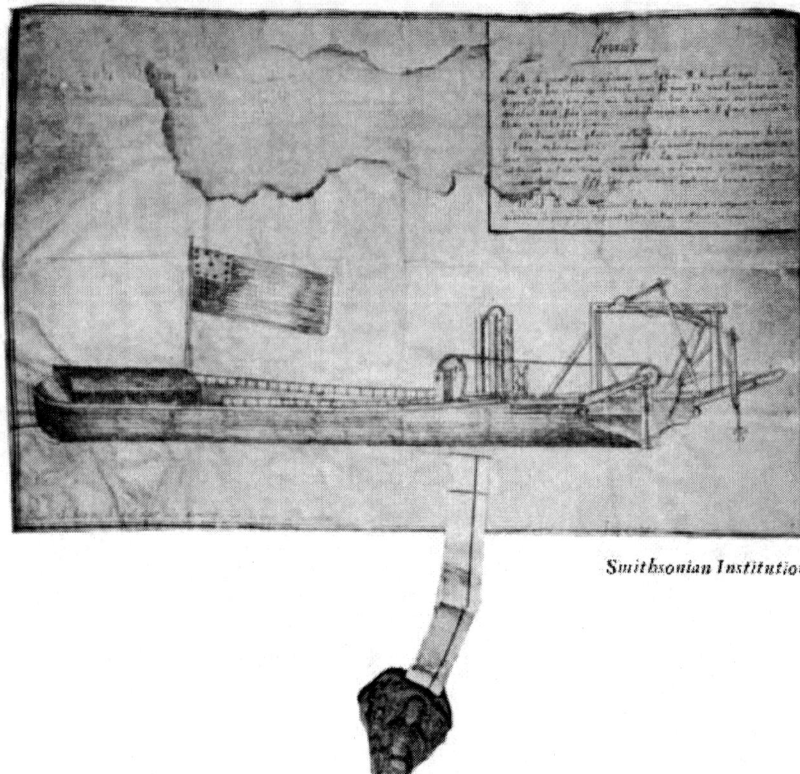

Smithsonian Institution

"POOR JOHN FITCH"

The patent application showing the means by which the American John Fitch hoped for a monopoly of steamship traffic in France

Smithsonian Institution

STEAM-DRIVEN GALLEY SLAVES

A model showing another extraordinary means by which John Fitch sought to navigate a steamship

associated then fell into a plebeian quarrel, accused him of bad faith, and refused to advance further money. D'Auxiron took them to court, obtained a judgment against them, and then, with this victory in hand, promptly died of apoplexy. The company was dissolved; with it dissolved the fifteen thousand francs which the ship had cost.

To salvage what they could, the heirs of D'Auxiron turned his designs over to the Marquis Jouffroy, who had belonged to D'Auxiron's original company. To him the King now transferred the option of a monopoly upon steam navigation should he be successful. Another company was formed and another ship was built by the mechanic Jacques Périer. There were trials on the river Seine in 1775, but the engines could not buck even a mild current and this company, too, was abandoned.

Thereupon Jouffroy retired to his country estate on the river Doubs. There, assisted by local talent in the form of the village blacksmith, he produced in June, 1776, a boat 14 feet long and of 6-foot beam, equipped with steam-driven duck's-foot paddles. This ship likewise proved aggressively unsatisfactory.

Still undaunted, Jouffroy pondered further and was shortly rewarded with a new and startling idea. The primary difficulty of applying a steam engine to marine navigation was that the steam engines of the day were reciprocating machines, but paddles had to be made to revolve. Because no one had really developed the crank-and-connecting-rod combination, the rock upon which every hopeful steam mariner struck was his inability to convert reciprocating into rotary motion without some extremely inefficient connection. But now Jouffroy was certain that he had the idea. He would build a new vessel 150 feet long, 16 feet in beam, displacing 150 tons, and having a draft of 3 feet. Into this he would put a steam engine with a horizontal piston. Now, thought

Jouffroy in high excitement, to the piston rod I shall attach a double rack with gear teeth upon its upper and lower faces—and between these gear teeth I shall set a pinion wheel upon the axis of the paddles. The paddles shall be 14 feet in diameter and shall dip 2 feet under the surface of the water, and I shall arrange so that the pinion wheel engages with the upper rack on the inward stroke and the lower rack on the outward stroke, thus making the pinion wheel revolve always in the same direction, regardless of the direction of the piston movement.

The engine and its rack connections were built at Lyons in 1780, by the Jean brothers. After another three years had elapsed, Jouffroy was ready for the first public trial of his craft, which took place on the Saône at Lyons on July 15, 1783. To the astonishment of everyone it seemed entirely successful, and a Proclamation to this effect was issued. But the Academy of Sciences, because the experiment had not been conducted in Paris where it held jurisdiction, withheld its indorsement. The government of France refused the monopoly for the same reason. Jouffroy could stand ridicule of his failures, but indifference to his success was too bitter a dose. Discouraged and disgusted, he took refuge in the army. France, having disacknowledged the locomotive and cast its inventor into jail twenty years before, was equally harsh to the inventor of a steam-driven ship. And with that, the curtain descends upon this European prelude, to rise again upon the first act in the soon-to-be United States of America.

One of the first men to continue the unfortunate experiments of William Henry was James Rumsey, who began his work in 1774. Casting out mechanical connections as clumsy and inefficient, Rumsey hit upon the same method of propulsion that was first suggested by Daniel Bernouilli and which is still being enthusiastically rediscovered every few years. As Sir Isaac Newton had done with a steam carriage, Rumsey proceeded on the reaction principle; he used an engine to drive a large pump which forced a stream of water from the rear of the boat and thus propelled the boat forward. In the days before cranks or marine propellers this method had the great virtue of simplicity. Indeed it was tried about a century later on a gunboat of moderate size by the British admiralty.

In 1786 Rumsey was credited with having driven a boat at four miles per hour against the current of the Potomac River at Shepherdstown, West Virginia, in the presence of George Washington. He received a patent for steam navigation from the state of Virginia in 1787, and in December of that year he is supposed to have repeated his earlier successes. But even then, despite the organization in Philadelphia of a "Rumsey Society" whose purpose was the encouragement of steam navigation, there were rumblings of dissension over his work.

Rumsey went to London in 1793 to supervise the building of a similar ship in England. The English were too much for him; on December 23rd he, like D'Auxiron, died of apoplexy while explaining his scheme to a London society. The boat which was then building was afterwards tried on the Thames, and is supposed to have traveled four miles in an hour. Yet with Rumsey's death went most of his claims, and by 1839 few remained loyal to him, even though a later Congress authorized President Van Buren to have a gold medal struck and presented to James Rumsey, Jr., in commemoration of what his father did toward "giving to the world the benefit of the steamboat."

When James Rumsey first began his work upon a steamship there lived in South Windsor, Connecticut, an eccentric, ingenious, and unfortunate mechanic. His name was John Fitch. Throughout his early years he lived a roving and not particularly distinguished career.

He was thirty-two years old when the Revolutionary War began. To his adventuresome and migratory type a war is always a pleasant interlude, and John Fitch soon found himself an armed rebel against King George III. He was a gunsmith and had the distinction of belonging to the shoeless heroes at Valley Forge.

The war having carried him from Connecticut to Pennsylvania, in Pennsylvania he stayed. And there in April, 1785, there occurred to him the same idea that was already agitating the minds of the Frenchman, Cugnot, and the Englishman, Murdoch—the idea of making the steam engine a means to land propulsion. But having had this idea, he straightway had another and, it seemed to him, more useful one; he would make the steam engine a means to water propulsion instead. Most of the world's commerce in the eighteenth century was borne on water; there the steam engine could serve a useful purpose, whereas on land, to John Fitch's eyes, it was likely to be little more than a curiosity. He began to build a model—and from that day until his death in 1798 he devoted his life to steam-propelled water craft.

With paddle wheels, and a small engine made of brass, Fitch tried out his first model on a stream near Davisville. It seemed to bear the promise of a great success, and when it was shown to Dr. John Ewing, provost of the University of Pennsylvania, he was so impressed with it that on August 20, 1785, he addressed a letter to one William Houston, warmly urging him to seek government aid for Fitch and his invention. Not long afterward Fitch, with a pouch of friendly letters, journeyed to the meeting place of the Continental Congress and there made application for financial assistance from the legislators of the new United States. These gentlemen had other affairs on their minds and were not in the mood to be interrupted by an impractical gentleman with a brass model of anything less than a federal constitution.

John Fitch was thanked warmly for his trouble in visiting the Congress and dispatched empty-handed with all possible speed. A madman, obviously.

Thus conventionally launched upon a successful career, Fitch redoubled his work. In September, 1785, he presented to the American Philosophical Society a model curious for the fact that the engine drove not paddle wheels but an endless chain of paddles strung between two axles in the fashion now made familiar by the caterpillar tractor.

In 1786 Fitch formed a company capitalized at $300. His partner was Henry Voight, a Dutch watchmaker. Fitch described him as a "plain Dutchman" but vowed that he preferred him to a Franklin, a Rittenhouse, and a number of other estimable people rolled into one. "He is a man of super-mechanical abilities," said Fitch, "and very considerable Natural Philosophy." Thus armed, the firm constructed a small engine having a one-inch cylinder. It cost some fifteen dollars and scarcely seemed worth it. But Fitch and Voight set to again and by August, 1786, they had finished a model boat powered by an engine with a three-inch cylinder. This time the endless chain of paddles was less appealing; by the exercise of "considerable Natural Philosophy" Fitch and Voight once again managed to avoid the use of a simple and sensible propulsion method. They connected the engine to oars—yet, after a fashion, the craft moved.

Fitch felt by now that he had prospered far enough to overcome the rebuffs of the Continental Congress. He made patent applications in a number of states (in those early days there was no central patent office) and in March, 1786, the open-handed state of New Jersey suddenly granted him a patent and a monopoly of steam navigation in that state for fourteen years. Pennsylvania at first was coy (having established its present character early) but in February, 1787, Delaware granted him a patent, and a month later New York did the same. Thereupon Pennsylvania fell in line. Fitch was now well protected. It remained for him and his "plain Dutchman" to build a steamship that would be an unquestionable success.

Now, happily, the effects of the Revolutionary War were beginning to recede. For one thing money was easier. So in August, 1786, Fitch and Voight began upon a boat 45 feet long and 12 feet in the beam, to be driven by a steam engine with a 12-inch cylinder. The trial of this brave ship came in May, 1787. The results are none too happy. The wooden cylinder heads leaked, the piston leaked, the valves leaked, the condenser proved itself entirely unequal to the simple task of changing steam back to water. The propulsion system was now even worse; the cylinder of the engine was connected to a crank axle by a walking beam; to the crank axle were fastened long rods projecting over the stern, with paddles lashed to the ends. As the axle revolved the paddles would dip down, push, rise from and return to the water in a queer halting cycle.

Fitch and Voight set about a remedy for the dreadful asthma of their craft. Voight designed an improved condenser, and one by one they overcame the sources of their earlier disappointment. Yet there were more troubles: the boiler now proved too small to feed the engine with its necessary calories. Fitch could say little more of this craft than that it made four miles per hour "with the tide." Since the tide of the Delaware River is somewhat speedy, this one of John Fitch's engines is still left with something to explain to its posterity.

At last, plied with further tonics for its frail constitution, the Fitch and Voight steamboat came to trials in Philadelphia on August 22, 1787. The members of the Convention, then in session at Philadelphia, busy in the framing of the federal constitution, saw it and were not excited. The Philadelphia newspapers were not interested; they minuted the fact that "a large green turtle has recently been caught in the Delaware," but unless they confused this with John Fitch's steamboat they made no mention of its voyage.

Whatever the success of the Fitch steamboat on that hot August day in eighteenth-century Philadelphia, it failed to impress even so open and flexible a mind as Benjamin Franklin's. He took one contemptuous look at it and decided that the capital expense, the layout for upkeep, and the amount of room in the hull preëmpted by the engine were all so large that steam navigation had not the faintest chance of competing with sails.

Fitch gathered his boat and the remnants of his dignity about him and went to work again. The year 1788 saw him finish a second full-sized craft, this time 60 feet long and 8 feet in beam. Again he increased the diameter of his cylinder, this time to 18 inches. Again he clung to his paddle propulsion at the stern. The boat was finished in July, 1788, and shortly afterwards distinguished itself by making a twenty-mile trip from Philadelphia to Burlington, New Jersey. The boiler expired at the last moment, but the craft was able to return to Philadelphia with the current. It made several subsequent trips at three or four miles per hour.

This was not a practical speed. But Fitch saw sufficient promise in the idea to warrant the construction of a third ship, which was finished in time to have her trials on April 16, 1790. She was a great improvement and traveled at seven miles per hour—a speed which no locomotive was to reach for almost a quarter of a century. Here, then, was the last word in transportation; the steamship belonged now in the service of commerce and communication.

> Although [said Fitch] the wind blew very fresh at the east, we reigned Lord High Admirals of the Delaware, and no boat on the river could hold with us.

In June, 1790, this ship was placed in the passenger run between Philadelphia and Trenton, stopping at Burlington, Bristol, Bordentown, and other way stations. In her simple, useful life she steamed, in all, two or three thousand miles, always avoiding the equally constant expectations that she would at any moment sink, catch fire, or blow up. No serious accident marred her blameless life.

So blameless it was that in the winter of 1790–91 John Fitch began his fourth steamboat—the *Perseverance,* obviously named by the inventor to give the public some idea of his bravery amid his sufferings. It was, however, never finished.

Fitch was now engaged in a long argument for a patent upon steam navigation from the United States government. He finally obtained it on August 26, 1791. Two years later, still flushed by this success and believing that all of steamboat navigation in the United States now belonged to him, he sailed to France in 1793, hoping to establish a similar monopoly there. He had not, alas, heard of the French Revolution. It was in luxuriant bloom when he reached France, and he was forced to work an ignominious passage home again in 1794. His star was waning now; he had had his idea of the steamboat earlier than the world showed any disposition to accept it and there was no money to reward him for his ingenuity. In 1796 he made one more effort to overcome his difficulty—the difficulty of an efficient propulsion method. He tried out a craft powered by a complete Archimedian screw at the stern (as opposed to the mere section of such a screw, which is the modern marine propeller) on Collect Pond in New York City—a site now enlivened by the Tombs, city prison of New York. But his boat was nothing more than a large rowboat, powered by steam, and it attracted no great enthusiasm. At last, beaten and discouraged, he settled in Bardstown, Kentucky, died there in 1798, and was buried, at his request, close to the banks of the Ohio River, soon to throng with the products of his neglected genius.

He had been a bitter man in his later years. His reiterated assertion had been that

> the day will come when some more powerful man will get fame and riches from my inventions; but no one will believe that poor John Fitch can do anything worthy of attention.

Both halves of that assertion to-day bear a sadly genuine, prophetic ring.

CHAPTER XIV

WITHOUT HELP OF NAPOLEON

SCOTLAND, the home of the steam engine, was not likely to stand idly by while upstarts in America, recently in rebellion, went ahead with plans to use this engine as a substitute for sails. Thus while poor John Fitch was struggling on the Delaware, similar girdings were taking place upon the Forth and Clyde Canal.

To Dallswinton, a wealthy Edinburgh banker had retired with ambitions to better the sailing vessel. He was Patrick Miller, who, in the 1780's, built a vessel "having rotary paddles in the two interspaces, driven by a crank." The crank was driven by men. It was hard work; when, one day, his son's tutor, James Taylor, took the briefest turn at the crank, he straightened up again and faced Patrick Miller —like the horns of Elf-land, faintly blowing. He insisted that another prime mover was needed.

Thus began a collaboration between Miller, Taylor, and one William Symmington, engineer. Quickly the combination came close to victory. Into a boat 25 feet long and 7 feet in beam, Symmington packed an engine having two 4-inch cylinders. Sensibly at last, this drove two paddle wheels, but in no sensible way. The paddles were not at the sides of the ship but, as with Miller's hand-cranked boat, in center-line wells, one fore and one aft of the machinery. They were connected to the engine not by the simple crank and connecting rod which everyone refused to discover, but by a fearful combination of weights, ratchets, pawls, and chains.

Yet the ship's first trial on October 14, 1788, went well; several hundred people on the banks of the Forth and Clyde Canal viewed what might be called a success. Patrick Miller was happy and spurred Symmington to further feats. In 1789 he finished a new ship—a twin hull vessel 60 feet long and driven by a steam engine with big 18-inch cylinders. It was tested in November and might have been more successful had not the paddle wheels disintegrated, parts of them floating gently upon the waters, leaving the willing engine helpless in the hull. December, however, saw a second trial, crowned with a seven-mile-per-hour achievement. But Miller had become dissatisfied with Symmington. He thought his expenditures excessive and his machinery too heavy; presently, abandoning an investment of £30,000, he washed his hands of Symmington and of steam navigation.

But Symmington's experiments had caught the interest of Lord Dundas of Kerse. It seemed to him that such a powered boat might make an efficient substitute for mules or horses to tow barges through the Forth and Clyde Canal. One cannot, however, hurry a peer of England. Lord Dundas pondered on the idea in his own good time and ten years later came to the point of engaging William Symmington to build him a new ship. He staked £7,000 upon the enterprise, and in 1801 Symmington went to work. A year later his craft was ready; in honor of its owner's daughter it was called the *Charlotte Dundas*—a patrician name for a plebeian tugboat.

By now Symmington had had the good sense to use a Watt double-acting engine with a horizontal cylinder, connected to the paddle-wheel shaft by a simple crank and connecting rod. Moreover, he had abandoned the silly idea of cutting a hole in the hull in which to insert the paddle wheels, and instead had placed them where they obviously belonged, on either side of the craft amidships. Thus it was that when, in

March, 1802, the *Charlotte Dundas* was brought to Lock Number 20 on the Forth and Clyde Canal, she took two vessels, each of 70 tons burden, in tow—and with Lord Dundas, Symmington, and others aboard, made Glasgow, 20 miles away, in 6 hours, bucking head winds all the way. Now, finally, it seemed there could be no question that a steamship could serve a useful purpose.

But the canal proprietors did not like this new idea. It was probably the sheer novelty of the thing which frightened them, but in their desire to cling to horses and mules they had one logical objection to the tugboat—that the wash of its paddle wheels would injure the banks of the canal. In this they were quite right. But even this did not daunt the enthusiasm of the Duke of Bridgewater, who quickly gave Symmington an order for eight boats to be used on *his* canal, backwash or no backwash. Then, having made his gesture, the Duke of Bridgewater died and the project was abandoned. The *Charlotte Dundas* was laid up in Lock Number 16, where she was allowed to rot, a monument to mixed precocity and prejudice.

But if, like the salmon, she had swum upstream only to die, she had also, like that same determined fish, seen to it that her race would endure. Among the witnesses to her triumph had been one Henry Bell, who, before long, was to build the first European passenger steam vessel, called the *Comet,* which in 1812 was to begin a regular route, traveling from Greenock to Glasgow on Monday, Wednesday, and Friday, and returning on Tuesday, Thursday, and Saturday —"four shillings for the best cabin and three shillings for the second." Even more important, the *Charlotte Dundas* had come to the notice of a young American whose name was Robert Fulton.

But Oliver Evans would like to be mentioned first. Evans was a Philadelphian who, not content with experiments on the locomotive and the steamship, has left behind him the inference that he invented the steam engine itself. He tells with touching modesty that in the year 1772 he sought "to discover some means of propelling land carriages, without employing animal power." But he had no idea what to use for energy until one of his brothers informed him

> on a Christmas evening, that he had that day been in company with a neighbouring blacksmith's boy, who, for amusement, had stopped up the touch-hole of a gun-barrel, then put into it about a jill of water, and rammed down a tight wadding; after which they put the breach-end of it into the smith's fire, when it discharged itself with as loud a crack as if it had been loaded with gunpowder. It immediately occurred to me that there was a power capable of propelling any waggon.

Here, in other words, was the old experiment of the Marquis of Worcester recurring on the eve of the American Revolutionary War. Evans continued with a story that later a book fell into his hands describing the atmospheric engine and that, shocked at its inefficiency in using the steam only to form a vacuum, he forthwith, single-handed, with no knowledge of James Watt, created all the necessary improvements.

His path, he admitted, was not easy. Benjamin H. Latrobe, the eminent federal architect, made life uncomfortable for him in his report to the Philosophical Society of Pennsylvania, and seems also to have made life uncomfortable for the steam engine by demonstrating that any attempt to make a steamboat was fore-doomed to disaster. Evans rode over his objections lightly. He tells of powering a sawmill with one of his engines, which had the faculty of using less and less fuel the more and more power it developed, "so that the more we resist the steam, the greater is the effect of the engine." Yet, although Oliver Evans may have been slightly more bumptious than he should, he did achieve one great success in 1804. For the Philadelphia Health Board he built a scow with a five-horsepower steam engine, to dig up mud and keep the channels open. Evans's blacksmith shop was a mile and a half from the Schuylkill River. To transport his scow that distance he mounted it upon wheels belted to the engine and, under its own power, drove it over the rough roads to the river edge, where he put it through one last trick—he made it launch itself. Thereupon he steamed with it down the Schuylkill and up the Delaware while the spirit of the vanished John Fitch gnashed its ghostly teeth, and Peter Cooper, being a thirteen-year-old boy, paid no attention whatever.

It was only after Oliver Evans left the scene that this idea of the steamboat began to grow at an accelerating pace. Even before his triumph the New York legislature in 1798 had granted to Chancellor Robert Livingston a state monopoly upon steam navigation, provided he could, within twelve months, produce a boat capable of traveling at four miles per hour. He failed. But in 1803 he procured a reënactment of the measure in favor of himself and Robert Fulton, who, having studied steam navigation in England, was now experimenting in France.

Fulton was born in 1765 at Little Britain, Lancaster County, Pennsylvania. He had had one of his earliest glimpses of the possibilities of steam when he visited the house of William Henry. He made youthful experiments with paddle wheels at the age of fourteen, when visiting an indulgent aunt who lived on the Conestoga River. As he began to grow up he learned the trade of watch-making; for, as parents of all famous men should be, his father and mother were bitterly poor and it was up to Robert Fulton to make his own way in the world. Unfortunately for them he early disentangled himself from a life among watch springs. In his early twenties, having decided that his bent lay in portraiture, he made his way to England, to study there with his fellow American, Benjamin West. He spent two years in Devonshire. There he met the same Duke of Bridgewater who was later to take up the success of Symmington's *Charlotte Dundas.* Under the various new influences which were reaching into his life he made a third shift in his career. Having abandoned watch-making for painting, he now abandoned painting for engineering. He went to France and there (he was moving in fine circles for a young man whose parents were too poor to send him to school) he met the Robert Livingston who had previously engaged in steam navigation and who was now his excellency, the United States ambassador to France. Together they determined to try a steamboat upon the surprised waters of the river Seine.

Fulton had learned his principles well, whether from Fitch or Symmington. On January 24, 1803, he delivered a model and a descriptive memoir to the MM. Molar, Bordel, and Montgolfier, who on L'Ile des Cygnes were building for Fulton and Livingston their first boat. With Fulton directing construction, the ship was finished in the spring of 1803. Now it was launched and Fulton smiled at it a beaming paternal smile. It smiled back, a trifle weakly, broke its back, and was immediately swallowed by the Seine. Its hull had been too weak for the heavy machinery it carried.

Fortunately that machinery was little damaged. The remnants were hauled out and repaired, and in July, 1803, Fulton and Livingston were ready again with a light ship, 66 feet long and of 8 foot beam. It made its trial trip on August 9th, and now, its spine properly stiffened, it cast loose from the dock and steamed up the Seine between banks lined by "an immense concourse of spectators." It made no better speed than any of its predecessors, but socially it made a great success. A committee (the MM. Bougainville, Bossuet, Carnot, and Périer of the National Academy) smiled upon it, and officers of Napoleon's staff lent the proper air of military importance. The boiler of this, the first steamship to be accepted in polite society, is still preserved in the Conservatoire des Arts et Métiers in Paris.

The happy Fulton now sought for his venture the blessing of Napoleon, then First Consul. It was an age in which inventors were becoming a nuisance. Not long ago the Italian, Alessandro Volta, had come to Napoleon's notice and Napoleon had bestowed medals and orders upon him. That was quite enough. Napoleon brushed Fulton's pleas aside and frowned importantly at the map of Austria.

But all was not lost. For what Napoleon had failed to supply, the legislature of the sovereign state of New York, U. S. A. (it is strange to think that these two existed in the same world), swiftly produced. Livingston asked for and obtained by letter a twenty-year monopoly on steam navigation, provided once again that within two years (later extended to four) Livingston and Fulton should build a boat that would sail the waters of the state at the proud speed of four miles per hour.

By now Fulton was in correspondence with the great firm of Boulton and Watt, steam engine manufacturers, and from them had ordered an engine with which his American craft was to be powered. In 1804 he went to England and from his plans ("From *my* plans," roared the outraged Symmington) saw Boulton and Watt build an engine with a cylinder two feet in diameter and a piston making a stroke four feet long. Then, under the sails which his success was so slowly to render obsolete, Fulton sailed for New York in October, 1806, and in December arrived home to begin building the hull of a vessel at first called *Katherine of Clermont*—Clermont being the name of Chancellor Livingston's country estate.

Science Museum, London

THE *CHARLOTTE DUNDAS*

MODEL of the first practical steam-driven craft to navigate Great Britain's waters

Science Museum, London

EARLY MARINE ENGINE

MODEL of William Symington's first design for propelling a tugboat

Smithsonian Institution

"FULTON'S FOLLY"

MODEL of the *Clermont*, first American steamship to win acclaim and confidence

Smithsonian Institution

COLONEL STEVENS'S *PHŒNIX*

MODEL of the first steam-propelled vessel to venture into open sea

In the spring of 1807 the *Clermont* was launched from the shipyard of Charles Brown, on the East River, New York. The hull was 133 feet long, with a beam of 13 feet. The depth of the ship, from keel to deck rail, was 9 feet. By August, the machinery shipped by Boulton and Watt had arrived safely and was installed.

New York continued in a becomingly incredulous pose. Even Fulton's friends could find not much excuse for him. He said afterwards:

> They were civil, but they were shy. Never did a single word of encouragement, or bright hope or a warm wish cross my path.

When it became known that Fulton was actually going to display his craft on the Hudson River by attempting a trip from the East River to New Jersey, a tremendous and incredulous crowd gathered to taunt him. The stage was well set for them. The silly boiler hissed and spluttered; the smokestack poured out fat clouds; during the difficult period of raising steam the nervous Fulton ran dementedly from one part of his craft to another. But steam was up presently, the lines were cast off, the throttle opened. Another roar went up when Mr. Watt's engine began to groan and yammer at its unaccustomed duty. Yet the ship moved, and moved smartly, and in a moment a certain change came over those who saw it. Fulton's friend, Cadwallader Colden, wrote:

> Nothing could exceed the surprise and admiration of all who witnessed the experiment. The minds of the most incredulous were changed in a few minutes. Before the boat had made the progress of a quarter of a mile, the greatest unbeliever . . . changed the expression of his features as the boat moved from the wharf and gained her speed; his complacent smile gradually stiffened into an expression of wonder. The jeers of the ignorant, who had neither sense nor feeling enough to suppress their contemptuous ridicule and rude jokes, were silenced for a

> moment by a vulgar astonishment, which deprived them of the power of utterance, till the triumph of genius extorted from the incredulous multitude that crowded the shores, shouts and acclamations of congratulation and applause.

It was one thing, however, to accept the steamboat as a *tour de force* and another to accept it as a vehicle. Thus when Fulton and Livingston announced that their craft could navigate the Mississippi and that, as a preliminary to this greater ambition, they would run it up the Hudson to Albany and return, the jeers broke out again. The Conservative Party thundered at Fulton's impudence; the clergy, no doubt, thundered at his impiety.

He persisted in it. For one thing, there was that twenty-year monopoly. On August 11th, with shouts of "God help you, Bobby" and "Bring us back a chip off the North Pole" ringing in his ears, he set off—smoke and flame of dry pine fuel pouring from the stack, the uncovered paddles churning froth and waves. He had one hundred and fifty miles to go, and through hostile territory. Farmers along the route, watching his progress on the hitherto undefiled Hudson, dashed home with the news for their womenfolk that the devil was sailing toward them "on a sawmill."

For this journey Fulton took the precaution of fitting the *Clermont* with sails, in case one of the fatal malaises of the day should seize Mr. Watt's engine. But none did. Leaving New York at 1:00 p.m. on August 11, 1807, the *Clermont* arrived twenty-four hours later at the home of Chancellor Livingston, 110 miles away. On August 13th, at nine in the morning, he cast off once again, and after eight hours covered the remaining distance of forty miles, reaching Albany at five that afternoon. The faithful engine had ground out an unremitting speed of five miles per hour. It had taken Fulton only thirty-two hours of running time to traverse the one hundred and fifty miles. Then, to pile triumph upon triumph, he returned to New York in a mere matter of thirty hours.

But the *Clermont*, although an engineering success, was no triumph of comfort or of safety. On the bosom of the placid Hudson she rolled like a gunboat in the Yangtze gorges. She had no berths in her cabin, no covering over her iron boilers—in short, none of the creature comforts which a river-going public was likely to demand. Furthermore she was unstable, for which reason she was rebuilt in the winter of 1807–08, and the beam widened to 18 feet. In the summer of 1808 she began a regular passenger service. Her passengers at first were few and frightened, but as her life continued unbroken by disaster, her popularity grew. By the end of the season she was too small for the crowds that sought her out. To the passengers of the day, used to sailing on sloops forever behind time, she gave much inconvenience, for Fulton insisted that she start regularly at her scheduled hour. Punctuality in transportation was something new.

The more nervous members of the Conservative Party, however, still continued to deplore her. She had very little appeal to the esthetic sense. She poured smoke into the fair sky, and she had a nasty habit of clanking. Dry pine was her fuel, and the flames it made rose above the top of her insufficient stack. She cast a trail of sparks behind her, and the crews of the more conventional vessels which she passed used to hurry below when she tore by in the night at her ungodly speed, "a monster moving on waters defying wind and tide, and breathing flames and smoke." Occasionally irate sailing masters would try to ram her, and the legislature was forced to come to her assistance by ruling that her destruction would be regarded as a public offense.

Of all the people who despised Robert Fulton and his steamship, none was more bitter than the gifted Colonel John Stevens of Hoboken. In 1789 he had petitioned the

New York legislature for a grant similar to Livingston's and had been refused. In 1804, when Fulton was in Europe, Stevens built a remarkable steamboat 68 feet long and 14 feet in beam, with a boiler of 100 two-inch water tubes to supply steam for a direct-action high-pressure engine, having a 10-inch cylinder and a two-foot stroke. Most remarkable of all, Colonel Stevens designed this ship to be driven, not by paddle wheels, but by a shaft carried through the stern, upon which there was mounted a four-bladed propeller—and this some thirty years before Ericsson brought the propeller into general use.

Then in 1807 Colonel Stevens built the *Phœnix,* a ship of paddle-wheel drive. To his scalding disappointment she was completed just too late to show her mettle before the *Clermont's* triumph slapped an iron-clad monopoly upon the waters of New York state. In June, 1808, the *Phœnix* was sent to Philadelphia under command of the colonel's son, Robert, who thereby made the first steam passage into open sea and weathered a storm to boot. Later these two broke down the monopoly of Fulton and Livingston in the courts, and thereafter their highly successful steamboats sailed the Hudson unmolested.

Despite the typical bitterness of two quarreling inventors, there was now plenty of glory in the steamboat for everyone. Fulton went on to further triumphs. He built the *Raritan* and the *Car of Neptune* (almost double the size of the *Clermont*) for passenger service. In 1812 he established the first steam ferry from New York to New Jersey, thus lending one of the earliest encouragements to that unhappy spectacle of modern life, the commuter. Each end of the ship was a bow, and each had a rudder. It was Fulton who contrived the floating docks which receive ferryboats to-day. The first steam ferry of them all traversed the distance of a mile and a half in fifteen minutes, and provided quarters for eight carriages, thirty horses, and some three or four hundred foot passengers.

The War of 1812 gave Fulton the opportunity of creating new boons for his countrymen. New York was frightened and Fulton, whose original "folly" was but five years old, offered to build a steam battleship for harbor protection. She was launched on October 29, 1814. With a lavishness which always characterizes navies, she cost some $320,000. She had provisions in her boiler room to heat shot red-hot for the greater discomfort of the enemy. Below the water line she had a submarine gun that could fire a hundred-pound shot. She had a greater displacement than the modern destroyer. But the War of 1812 was inconsiderately concluded before she was ready for action.

Fulton himself did not see this really formidable warship completed. In January, 1815, he had traveled the tiresome journey to Trenton and there had worn himself down by wrangling with the New Jersey legislature over the operation of another steamship line. Coming from the heat of debate into the cold of winter, he took a chill, and on February 24th, at the age of fifty, he died. But the flames still roared in the smokestacks of his little vessels on the Hudson, and every year thereafter they were to light the dark waters of the sea in a broader and broader circle.

CHAPTER XV

SPOKEN AT SEA

WHEN Captain Moses Rogers gave his orders to cast off in the harbor of Savannah, Georgia, on a late day in May, 1819, it became evident to the townspeople that his insanity was not of the temporary sort they had hoped. The orders were obeyed by his equally mad crew, and the ship *Savannah* dwindled away in the roadstead while the townsfolk went home to discuss his folly at greater length. Savannah was a prosperous port in those days, and there was nothing unusual in the departure of a ship on a transatlantic voyage. But there was something very unusual about Captain Rogers and about this particular ship. The captain, so came the rumor from the North, had been associated with some queer marine experiments; he had commanded two vessels known as the *Clermont* and the *Phœnix,* and now he had gone to sea bound for Liverpool, England, in a wooden ship that had a fire burning in the middle of it to make steam for an engine that was to turn paddle wheels. A steam engine in mid-ocean? It was unthinkable.

This demented ship had been built in the North by Messrs. Crocker and Fickett at Corlears Hook, New York. The eccentric Mr. Scarborough of Savannah had purchased her, secured the services of Captain Rogers (who likewise had a hand in her building), and sent him off upon his insane journey.

Of course, it might have been worse. The *Savannah* was a full-rigged ship and was therefore not at the mercy of steam. Yet in the center of her she did have something that was referred to as an inclined, direct-acting engine, with a cylinder 40 inches in diameter and a 6-foot stroke. It was designed for ten pounds of pressure to the square inch, but would feign a little honest work on two. Those connected with the voyage insisted that if enough fuel were ignited under her boiler this engine would develop the power of 90 horses and would thereby drive two wrought-iron paddle wheels.

The *Savannah* had been launched on August 22, 1818, had sailed from New York for Savannah on April 27, 1819, reaching there in seven days. Now, on May 26th, she was gone again, bound for Liverpool. Very well; there had been no deterring her master, so let his ship lose herself somewhere in mid-Atlantic and let Captain Rogers suffer the unpleasantness of death by immersion and exposure.

Now it is quite true that Captain Rogers was destined shortly thereafter to die. It is also true that the *Savannah* was to be wrecked. But neither of these misfortunes bore out the predictions of the Savannah townsfolk. For when Captain Rogers passed away, the nemesis that destroyed him was not drowning but a mosquito bite. He died in 1822, a victim of yellow fever. And although the *Savannah* was to be wrecked, she was first to travel from Savannah to Liverpool, from Liverpool to St. Petersburg, Russia, and so home, without disaster. It was long afterward that she met her fate; running as a packet boat, she gashed herself to death on the shores of Long Island in the same year that Captain Rogers died.

But the crossing of the Atlantic by the *Savannah* was not an unmixed triumph for steam. Even Captain Rogers had incomplete confidence in its power. On his little ship of 350 tons burden he had so arranged the paddle wheels that they could be taken in-board and lashed on deck whenever the seas became rough enough to menace them. Thus it was that on a voyage lasting twenty-nine days steam was raised only six times and the engines were run a total of only eighty hours. Even so, the entire cargo space of the *Savannah* was insufficient to store enough coal and wood for its greedy and wasteful boiler.

The *Savannah's* arrival in Europe was somewhat confusing to the Irish. On June 17th she had lowered her sails and, with smoke pouring from her single funnel, she was steaming off the coast of Ireland. To the Irish, smoke on a ship meant only that she was afire. Coastguards of the Cape Clear Station so reported, and the cutter *Kite* was dispatched after her in relief. Yet to the amazement of the *Kite's* crew their trim cutter could not catch up with a ship running under bare poles. When an Irishman is mystified he fights; the *Kite* immediately fired several shots at the *Savannah,* and Captain Rogers (he was chief engineer as well as master) stopped his engine. The crew of the *Kite* boarded this demon ship, gazed in awe and wonder at her power plant, apologized warmly, and let her go. She reached Liverpool three days later.

To her first official port of call she brought neither passengers nor cargo, for she had left Savannah innocent of both. No passenger would trust himself to a ship with a fire in her belly; shippers of freight were equally coy with their cargoes. But when the *Savannah* left Liverpool on July 23d, bound for Copenhagen, Stockholm, and St. Petersburg, she carried with her a passenger, the brave Lord Lynedoch, who, when he arrived safely at St. Petersburg, gratefully presented to Captain Rogers a silver tea kettle, suitably inscribed. It is now too late to say whether he intended his gift to be used for making tea or as a substitute for the *Savannah's* boiler, which had completely used up seventy-five tons of coal and twenty-five cords of wood during eighty hours of making steam.

Thus was the Atlantic first crossed by steam. But it had been an uncertain victory, and when Captain Rogers returned to the United States in December, 1819, after a round trip that had lasted more than six months, he did so completely under sail, indignantly refusing to pay European prices for fuel. The *Savannah's* paddle wheels remained lashed under canvas covers upon the deck. Once again the ship had been empty of passengers. Her machinery was ignominiously removed and thereafter she lived and died a sailing ship.

It was to be almost half a century before another steamship of American registry was to venture across the Atlantic. To the backers of the *Savannah* it seemed that they had spent $50,000 upon a foolish venture. The Conservative Party not only decided that transatlantic steamships were impractical, but adduced the best of scientific authority to back them up. A Harvard professor of mathematics, casually assuming that engine efficiencies would never improve, showed that it was impossible for any vessel ever to cross the ocean propelled only by steam, since no ship could ever carry sufficient fuel.

Everyone believed him until the year 1838, notwithstanding that in 1833 a steamship called the *Royal William* had trudged across the ocean from Canada. But in 1838 the voyage was twice accomplished by English ships carrying auxiliary sail. The *Sirius,* of 700 tons burden and 250 horsepower, sailed from Cork on April 4, 1838, and the *Great Western,* of 1,340 tons burden and 450 horsepower, sailed from Bristol on April 8th. Both arrived in New York harbor on April 23d, the *Sirius* in the morning, the *Great Western* in the afternoon. The *Great Western* had burned 660 tons of coal to drive her 28-foot paddle wheels. Seven passengers had dared to make the trip with her. In point of time, the distance between Europe and America had now been cut in half. Forts and men-of-war in New York Harbor thundered their salutes, merchant vessels dipped their flags, and the excited citizens of New York assembled on the Battery to see the new day ushered in. Each vessel left for home again a few days later, whereupon the *Sirius*, being considered too small for so boisterous a route, crossed the ocean no more. But the *Great Western*, which was all of 236 feet long, stayed in the trade and made seventy passages in the six years that followed.

Capital took a hand now and one Samuel Cunard brought his ship *Britannia* into New York Harbor in the summer of 1840. Soon, by his success, he accumulated a fleet of four vessels—a total tonnage of 4,600. The ships had "side-lever" engines which would drive them at speeds a little less than eight knots. The owners of the great American clipper ships that ruled the commerce of the ocean were properly contemptuous. A stolid, trudging steamship could never compete against these thoroughbred beauties, with their stun'sails, clipper bows, spankers, and dolphin-strikers. Considering the weak, simpering horsepower with which steam vessels up to then were equipped, there seemed some reason for this toplofty attitude. Marine engineering had bad need for another genius.

It was soon to discover him in the person of John Ericsson, who was born in Sweden on July 31, 1803—at about the time when Richard Trevithick was experimenting with his locomotives.

Ericsson spent some of his early years in the Swedish Army. He found time to indulge his interest in machinery. Having developed a prime mover which he called a "flame engine," he put military life behind him and journeyed to England to find encouragement. He found little. But he did find one John Braithwaite, with the result that the firm of

Braithwaite & Ericsson was soon formed to concern itself with pumps and boilers as mine equipment.

One of the earliest important commissions which this new firm undertook (not forgetting the Rainhill Contest for locomotives) was the construction of engines for the ship *Victory,* used by Sir John Ross between the years 1829 and 1833, in his second attempt to discover the Northwest Passage. Arctic winters are hard upon thermal efficiencies and human temperaments. For these reasons Sir John came soon to believe that he valued the room taken up by the engine more than he valued the engine. Being a man of action, he ripped it out, and flung it overboard, where perhaps it lies to-day, waiting to astound an archeologist of the twenty-fifth century, who will discover that Eskimos invented it. After this vandalous gesture, Sir John dusted his hands and said, "I believe there was not a man who ever again wished to see its minutest fragment." When he returned home with the Northwest Passage but without the engine, John Ericcson, who from birth had had a choleric disposition, charged the discoverer "with forgetfulness of justice and candor," and was saved for science only by friends who persuaded the two not to fight a duel.

Ericsson did not like the paddle wheel, whose inabilities to cope with waves bothered him. For centuries there had lain ready at hand a device to take its place, a device invented by Leonardo da Vinci, and known as the aërial screw. But few people had successfully put this curiosity to work at sea. John Ericsson determined to do it.

Some time before 1833 Ericsson had been called upon by the London Carrying Company to experiment with submerged propellers on the London and Birmingham Canal. By 1835 he had designed a propeller to be driven by steam and which consisted of a series of segments of a screw attached to a hoop, supported by arms so twisted that they formed part of a screw somewhat after the manner of the modern windmill. Encouraged, in 1836 he designed a steamer to be driven by two screws, one behind the other, the outer one being turned by one shaft which ran inside the other—for Ericsson thought that the turning effect of a single propeller might swerve the ship from a straight course.

Ericsson had a warm friend in the person of Francis B. Ogden, who, coming from the state of New Jersey, U. S. A., happened to be American consul in Liverpool. It was he who, seeing Ericsson's plans and the value that lay in them, advanced part of the money with which a new ship was built, to be launched on the Thames in 1837. The *Francis B. Ogden* proved capable of traveling ten miles per hour and was therefore known as "The Flying Devil." Used as a tug, she towed a schooner of 140 tons burden at seven miles per hour. Here, thought Ericsson, is an achievement which might interest the British admiralty. He arranged a demonstration, but the admiralty yawned behind its palm and left for tea as soon as possible. Engine builders were likewise opposed to the screw, which meant the redesigning of all their machinery. Having contributed the steam engine to the world, they now felt that the world was unimprovable.

In that year still another man from New Jersey was in London. He was Captain Robert F. Stockton, U. S. N., who was then building the Delaware and Raritan Canal in the United States. Having invested all his family fortune in it, he had come to England to raise additional money. Through Ogden, Captain Stockton met John Ericsson, liked him and his ideas, and suggested that there was more opportunity for an ambitious young man in the United States. The result was that when the steamship *Great Western* sailed for New York on November 1, 1839, she carried John Ericsson, resolved now to introduce the propeller into the United States.

Ericsson, when he came to America, was thirty-six years

old. He was a large, fiery, powerful, and determined man, who, alas for the statistics of the Women's Christian Temperance Union, drank enthusiastically and continuously up to the age of fifty, when, with equal determination, he stopped altogether. His wife later followed him to the new country. She took one look at it and one at her husband. Shortly thereafter she returned to England, declaring that her husband was "wedded to a steam engine" and had no need of her. They never met again.

Fortunately for Ericsson's marine engineering plans, Congress, in 1839, had authorized three war vessels for the navy. Two had immediately been ordered, one of them the *Princeton,* and Ericsson had been commissioned to build the machinery. He constructed a peculiar set of engines of the "semi-cylinder" type first designed by James Watt. Each engine was composed of a vibrating rectangular piston, moving inside a semi-cylinder. Ericsson connected two of these machines at right angles to the shaft; and on the shaft he placed a six-bladed fourteen-foot bronze propeller, so installed that it could be disconnected from the shaft and hauled up into a cavity in the ship's hull—thus avoiding a serious drag when the ship was under sail. Thus he produced a ship capable of traveling thirteen knots—perhaps the fastest warship afloat in 1844. Captain Stockton commanded her, and a great fuss was made over her by navy officials. She ill repaid them. On February 28, 1844, one of her biggest guns burst, killing, among half a dozen people, two officers of President Tyler's cabinet, who were inspecting their darling.

Nevertheless, the unlucky *Princeton* had brought about another revolution in the application of power. Early engineers despised the propeller, for it meant a complete alteration of the types of engines that must be built for it. The slow-speed engine which was sufficient for a paddle wheel must now be supplanted by a high-speed engine—for it was necessary to drive a propeller at from two to three times the revolutions per minute of the paddle wheel.

Despite objections, the propeller would not be denied. In 1841 six ships had been fitted with John Ericsson's development. By 1842 there were nine. But by 1843 there were about thirty. The die-hards among marine engineers protested that there was no place to put a propeller where it would not interfere with the rudder; that the idea of punching a hole in the stern of a ship for a shaft to pass through was dangerous and immoral, and that anyway there was no material for shaft bearings which would not wear so badly that water would escape past it and sink the ship, unless possibly lignum-vitæ blocks were used, in which case, they insisted, these would catch fire from the friction. The problems were certainly difficult. So were those of proper size, shape, pitch, and immersion of the propeller—so difficult that marine engineers are still wrangling about them to-day.

The propeller's great protagonist fought for his idea unceasingly. He was an aggressive and opinionated man who had no hesitation in describing himself in a letter to a friend as

> . . . the person who has done more to promote marine engineering, mechanical motors, implements of naval warfare, etc., than any other ten persons together during the last third of a lifetime of the American Republic.

In this he was perhaps quite right, if not quite modest. His contributions to marine engineering were tremendous. He devised an artificial forced draft to make for better steaming; he improved the surface condenser; he devised the hydraulic reversing gear, and a feed water heater. He shouted the advantages of high-pressure steam.

But the Ericsson of these ideas, of the battleship *Princeton,* of an ingenious hot-air engine, and of the historic Rainhill Contest for locomotives, is all but forgotten, whereas the John Ericsson who built the *Monitor* for the Union forces during the American Civil War still marches on through every elementary history book in the land. The *Monitor* was a great achievement, but many others in Ericsson's life were more noteworthy.

One of Ericsson's few errors of conservatism lay in his damnation of the compound or multi-expansion engine, which, he asserted, was merely a device of English engine builders to increase the cost of their products. Ericsson could see that the steam engine was still an uneconomical machine, and he had hopes of his own to "materially benefit civilization by checking the rapid exhaustion of the coal fields which threatens to put a stop to human progress in the not distant future." Presumably, then, it was only through stubbornness that Ericsson denied the triple-expansion engine. In marine power, despite him, it swiftly took its place, as the advantages it offered were discovered. The volume, temperature, and pressure of steam are all intimately related; economy lies in putting steam into the engine as hot as possible and taking it out as cold as possible. A housewife always extracts a little more dirt every time she rinses her washing; just so engineers can rinse a little more energy from steam by putting it through more than one process of expansion. The trick, then, that occurred first to Jonathan Hornblower (viciously attacked by Watt for his pains) and later to Arthur Woolf was not to release the steam into the condenser after it had done its original work but to release it instead into another cylinder which, designed to operate at a lower pressure, would wring some additional energy from the steam's expansion. Efficient engines to-day do not merely use double expansion; they use triple and occasionally quadruple expansion. Such an arrangement is vastly more efficient, but the large, bulky, and complex machinery can seldom be used upon such crude power plants as locomotives. The multi-stage engine was soon to have great application in marine engineering, and it is heartily too bad that John Ericsson opposed it with all the force of his powerful personality.

In the latter part of his life the testy Ericsson lived at 36 Beach Street, New York City, in a neighborhood which, to his complete satisfaction, degenerated rapidly during his ownership. He had an intense dislike for most of his fellow men, and the less he was bothered by social affairs, the happier he became. Even when, in 1866, he was scheduled to receive the Rumford Medal (voted to him four years before), he wrote at the last moment to Charles Gould, one of the committee on arrangements:

> Apart from my utter repugnance to the proposed ceremony, I yesterday had the misfortune to hurt my back so seriously as to bring an old complaint originating in lifting a heavy weight. Many weeks will elapse before I can get out.

And to Professor Eben Horsford, Rumford Professor of Chemistry at Harvard, who was to present him with the medal, he said with a somewhat closer approximation of the real reason,

> . . . the recipient of your favors is an eccentric person, who nothing can induce to appear in public, and who has not been out of his house for five years. . . .

Through his eccentricities, John Ericsson did yeoman service in noise abatement, even in his lifetime a serious drawback to a happy life in New York City. With the freight cars and engines across the street which steam had brought into the world, and with the drays which still competed with them, he could do but little. But with crowing roosters (one of the few sources of noise to have disappeared from Manhattan Island) he could do a great deal. He sought them out enthusiastically, purchased them from their owners and then,

SIDE-LEVER ENGINE
FOR years this type of machine drove ships on the high seas

Smithsonian Institution

STEVENS'S ENGINE
THE precocious product of Colonel John Stevens, designed in 1804 to drive twin propellers

Baltimore & Ohio Railroad Archives

OLIVER EVANS'S SCOW

A MODEL of the first steam-driven amphibian in history which trundled itself on land to its own christening

Smithsonian Institution

EARLY ERICSSON

MODEL of the *R. F. Stockton,* showing Ericsson's early conception of the propeller

in a high religious zeal, wrung their necks. He paid a neighbor five dollars to take the neighbor's dog somewhere else to bark for a period of a year. On another occasion he furnished the money to pad a neighbor's wall with mattresses—tastefully covered to match the furniture—so that the sound of a piano from the opposite side of the wall would not disturb him while he worked. A further service to civilization he performed by presenting two handsome gold watches to the same number of little girls, upon the understanding that they would thereafter desist forever from practising their scales within hearing of his rooms.

His engineering interests never brought him to the point of riding upon the puffing, chugging New York elevated railways. He never in his life saw Central Park. He traversed Brooklyn Bridge only when tricked into it by a conspiring secretary, and for a long while refused to believe that such an instrument as the telephone existed even when he heard a voice transmitted by it. At the age of sixty he used still to stand on his head to demonstrate his athletic ability, and never during the daytime would he lie down on a bed—preferring, in a more manly fashion, to take daytime naps on the top of a table. Winter and summer he wore vests and stocks made of a buff piqué. The pattern had pleased him and he bought one hundred and fifty yards of it to use for these garments all the rest of his life. He had the same housekeeper for thirty years of his life, part of whose ritual it was to tuck him into bed and say good-night to him at the close of every day.

He died on March 8, 1889, on the twenty-seventh anniversary of the battle of the *Monitor* and the *Merrimac*. His housekeeper, when first he had begun to fail, had bought him a reclining chair, but as life left him, his voice was still muttering feebly against it. It was made, he said, upon a false mechanical principle.

CHAPTER XVI

Steel Shoulders

"The modern fashion in all things, is to 'go ahead,' push on, keep moving, and the faster the better—never mind comfort or security or pleasure. Dash away, annihilate space by springing at a single jump, as it were, from town to town, whether you have pressing business or not." It has the modern sound, this wail for the vanished past, but it was written by Samuel Breck in his diary for December 31, 1839.

True enough, railroad patrons in those early days were transported from place to place upon something notably less luxurious than flowery beds of ease. The same diarist had a few more acid remarks to make which made his objections crystal clear:

> "How do you mean to travel?" asks neighbor John. "By railway, to be sure, which is the only way of traveling now; and if one could stop when one wanted, and if one were not locked up in a box, with fifty or sixty tobacco chewers; and the engine and fire did not burn holes in one's clothes; and the springs and hinges did not make such a racket; and the smell of the smoke and oil and of the chimney did not poison one; and if one could see the country and were not in danger of being blown sky high or knocked off the rail—it would be the perfection of traveling."

Yet somehow the railroads progressed. By 1833 the Baltimore & Ohio had reached Harper's Ferry and with its 137 miles of track was the longest railroad in the world. On August 25, 1834, it made its triumphal entry into the capital in Washington. It was working hard to make travel more civilized. Already there had been developed an eight-wheeled passenger car (the first was called the "Comet") which, abandoning the early tradition of the stagecoach, had forty seats ranged on either side of a central aisle. The seats were cushioned and reversible in direction.

Moreover, the federal government made an important gesture. Postmaster General Kendall ordered that thereafter mail from Washington to Baltimore and Harper's Ferry be carried by railroad instead of by stagecoach. Thus, in 1834, the first mail car traveled over rails in the United States. Actually it was nothing more than one end of a baggage car, boarded up. There were no railway mail clerks; the compartment was fitted with a lock to which only the postmasters in Washington and Baltimore had keys. One postmaster locked up the mail; the other unlocked it.

Happy to have wooed and won their government, railroaders turned their attention now toward new refinements, particularly in that one vital part, the locomotive. By 1836 the American engineer, William Norris, had built a locomotive, the "George Washington," to astound the world. It had six wheels, two of which were four-foot drivers. The firebox was placed well forward and was partially carried by a four-wheeled truck in front. The whole weighed 14,400 pounds; with its outside cylinders it soon performed the supposedly impossible feat of pulling 19,200 pounds at fifteen miles per hour on an incline that rose one foot in every fourteen.

This performance had one intensely significant result; for the first time locomotives crossed the Atlantic Ocean from west to east. The United States, hitherto an importer, had now become an exporter of railway motive power—and this only eleven years after England had developed the first steam railroad. By 1840 Norris had delivered the "Philadelphia," "Columbia," "Atlantic," and "England" to the Birmingham & Gloucester Railway in England—the first of an order for eight locomotives built upon the plan of the "George Washington" and weighing some nine and one half tons. Of these, the "Philadelphia" was selected to show its strength by climbing the "Lickey Incline" up which trains had previously been pulled by a winch and rope. This grade was two miles long and had a rise of approximately one foot in thirty-seven. It was believed impossible that a locomotive could draw a train to this Olympian summit, but the "Philadelphia" did so easily—whereupon Norris, to an obbligato of gnashing English teeth, received an order for sixteen more of his estimable locomotives. The Birmingham & Gloucester Railway abandoned fixed machinery at grades, and the stock of the enterprise moved up by some five pounds per share.

Like most animal organisms, the locomotive reached its greatest speed before becoming endowed with its greatest strength. The loads that locomotives could pull in the first half of the nineteenth century were inconsequential by the standard of to-day, yet the century had not half gone by before speed had reached levels as high as are in force to-day. Herein, perhaps, lay the greatest change in the habits of mankind which had ever come over the race since it emerged from slavery; man could now travel, in one hour, a greater distance than, in a steamless epoch, he could travel in a day.

Another great step which America made with its new seven-league boots was taken by Ross Winans in 1843. Winans built the "Carroll of Carrollton" in remembrance of that Charles Carroll who had shoveled away the first load of earth that impeded the progress of the Baltimore & Ohio Railroad. Like the colt that is all legs, the "Carroll of Carrollton" was all drivers, and when it came to trial upon the Washington branch of the Baltimore & Ohio Railroad it became evident that no one dared open the throttle wide. Roadbeds in 1843 were not in condition to make high speed anything but a daredevil chance.

Thus the "Carroll of Carrollton" brought American railroad builders abruptly to realize that it was now not the locomotive but the roadbed which limited speeds. Hitherto every scrap of rail used in the United States had been imported from England. It was not until 1844 that the first rail was rolled in the United States. Mount Savage, in Allegany County, Maryland, saw this historic event take place, and the Franklin Institute in Philadelphia struck a silver medal in its honor.

But the home of railroading did not look lightly upon the attempts of the United States to make itself technically independent. English locomotive designers still had a few tricks up their sleeves—as the "Great Western," built by Gooch in 1846, testified. The locomotive now was no longer a vertical engine set upon four wheels and cranking away at them through the medium of a walking beam. It was beginning to assume the general type and appearance of the locomotives of to-day; a stack at the front, two or four small wheels in advance of the drivers, and horizontal cylinders fed from a horizontal boiler. Boilers were invariably tubular, and the engine driver had taken up the conventional position at the rear of the machine. There remained two main points upon which designers were still unsettled: the fuel was as often wood as coal; and the cylinders were as often placed *inside* the drivers as *outside*.

The "Great Western" had inside cylinders but in other respects she approached modernity. She achieved, in the year 1846, a maximum speed of 78 miles per hour, and was fed by a boiler whose heating surface totaled 1,800 square feet. She found it possible to start from Paddington and arrive at Didcot (53¼ miles away) in 47 minutes—a speed that approaches the highest scheduled speed now attained on any railroad in the United States.

Obviously the steam engine which reached such velocities

was an improved mechanism over the simple barbaric machines which had participated in the Rainhill Contest but seventeen years earlier. And improved it was, in one intensely important particular—the valve gear.

Early engineers had found it no small problem to devise a mechanism by which, upon a moving vehicle, cut-off might be varied from the long steam admission necessary in starting to the shorter which was economical at higher speeds. Moreover, a locomotive which was not reversible was a sadly limited affair.

The valve gear which admitted, cut off, and exhausted steam from the double-ended cylinders of early locomotives was a rudimentary apparatus. It consisted of loose eccentrics upon the driving axle made to engage with forward- or backward-driving studs by means of a treadle under the control of the engineer. With this arrangement speeds were limited not so much by boilers as by the rack and pound of the crude mechanism.

But by 1841 there had finally been developed a device known (with no particular justice, but with a unanimity which still earmarks it to-day) as the Stephenson Link. By means of this, fixed eccentrics fastened to the driving axles drove a curved rocker arm in a leisurely back-and-forth movement. The rocker arm was slotted, and in this slot there could move from top to bottom (as controlled by a lever available to the engineer) a block of metal by which the rod connected to the valve was driven. When the metal block was at the extreme top of the rocker-arm slot the valve would open to its fullest width and stay open its longest time. As the metal block was moved, by the action of the engine driver, closer and closer to the center of the rocker arm, the motion of the valve became less and less, thereby producing a shorter and shorter cut-off until, in the exact center of the rocker arm, the valve moved not at all and the engine was out of gear. Below the center of the arm the valve put the engine in reverse—for a reason analogous to that whereby the seasons are reversed for anyone who travels from the northern to the southern hemisphere.

Thus, by one great invention, perhaps the greatest since James Watt had devised the separate condenser, it became possible for the engine driver to vary steam cut-off for the enhancement of engine efficiency and to use the same mechanism to put his machine into reverse. Furthermore, the Stephenson link was a positive mechanism with all parts constantly engaged.

As was true of so many other locomotive developments, the Stephensons, father and son, were none too clearly entitled to credit for the link. It was William Williams, a Stephenson draftsman, who had the idea for the link, and William Howe, pattern maker, who made it workable.

The Stephenson link-motion was to endure for some seventy-five years; even in the present day it continues to actuate the valves upon small passenger locomotives and some larger ones. It was not generally superseded until the comparatively recent and more complicated motions which go by the name of Walschaerts, Baker, Southern, and so on. Such an enduring mechanism was, to the generation which saw its development, an imperishably permanent addition to locomotive design; fortified by the genius which had produced it, the locomotive emerged from its juvenile into its adult period. Henceforth its development was to be complex and technical; the simple day of the mechanism had passed by the year 1850.

In this same year the United States had some 8,600 miles of railway—built at a cost of almost $300,000,000 and usually connecting cities along the Atlantic Coast. Within three years the Baltimore & Ohio Railroad was to reach its original objective of Wheeling, West Virginia, and thus make itself known as the longest railroad in the world—379 miles between ultimate terminals, and built at a cost of $15,500,000.

Even then, the railroads had done a puny job of linking the parts of a nation together. By 1850, sixteen steam-driven ships were plying the Sacramento River, but California had yet to hear her first locomotive whistle. Upon all the West Coast of the United States there was no such thing as a self-propelling steam engine, running upon rails.

It was some time before this lamentable state of affairs was redressed. At last, however, the entire section west of the Rocky Mountains was treated, through the pioneering genius of one man, to its first locomotive. The Sacramento Valley Railway came into being, running between Sacramento and Folsom, twenty-two miles away.

It was the first frail link in the gigantic chain which the imagination of Theodore D. Judah had forged in his mind. In 1860 Judah had run a survey over the Sierra Nevada Mountains with the idea that, although all practicalities seemed against it, a railroad connecting with the East (an indefinite locality stretching somewhere between Nevada and Rhode Island) was essential to the development of California. Ox teams took some six months to span the continent, and for all the bellowings of politicians, a nation was not, under those circumstances, a nation at all.

But for its economic salvation California was unable to furnish money. Judah came to Washington to lay his Pacific Railway scheme before Congress. Congress took it as a God-given subject for debate but was guiltless of action. Soon, however, Judah met those four strong men, Leland Stanford, Mark Hopkins, Charles Crocker, and Collis P. Huntington, and with their aid the Central Pacific Railroad was incorporated on January 28, 1861, with a capital of $8,500,000, of which the modest fraction of $150,000 was paid in.

Judah returned now to Washington, and with this backing won the day. Congress passed its historic act authorizing the construction of the first continental road. It further specified:

> The track upon the entire line of the railroad and branches shall be of uniform width to be determined by the President of the United States, so that, when completed, cars can be run from the Missouri River to the Pacific Coast.

The question of gauge was difficult and dangerous. In the South, which had mothered the railroad, the gauge was 5 feet. In the East it varied from 4 feet 8½ inches to 6 feet. The state of Missouri had its own gauge, which was 5½ feet.

The problem was now in the lap of President Lincoln. Incredibly, he decided that the proper gauge was 3 feet. Judah and the state of California, where the gauge was 5 feet, rose in protest. Lincoln changed his mind, and said that 5 feet should be the legal gauge for the United States. It was a change in the right direction. A 3-foot gauge, had it been accepted, could quite well have throttled the Union at the very instant when Lincoln believed he was delivering it; a matter of inches was to determine whether railroads could be sufficiently swift, commodious, and powerful to span a continent three thousand miles wide—or whether a meaner conception of travel was to make East and West as separate as Denmark from Peru.

Yet still the problem was not settled. There were powerful interests which bitterly fought this idea of presidential sanction for gauge. To placate them shortly Congress passed a new act repealing the presidential prerogative and once and for all settling the gauge for the railroad from Missouri to the Pacific as 4 feet 8½ inches. The matter might have had a happier outcome, but at least it had an outcome and that was something.

Among the many provisions of the Act of Congress there was one which specified that the Central Pacific, the railroad in which Judah was concerned, should build but one half the right-of-way, working eastward, while the Union Pacific Railroad, working from Missouri, would travel westward to meet it. On January 16, 1863, to the accompaniment of blaring bands, Leland Stanford in his capacity as governor of California thrust in the first shovel and the great work began. The celebration was marred by the death, a few days before, of Judah, the driving spirit of the enterprise. "Panama fever" struck him down at the age of thirty-seven, but the vast work went on in the face of desperate hardships for the workers and towering skepticism from the sidelines. The eastern portion of the road was the easier to build and progressed the faster. The western portion, which had the incredible task of crossing the Rocky Mountains, toiled ahead as an army of twelve thousand men slowly spun out a double iron thread behind them. Finally the two forces sighted one another and on May 10, 1869, the two roads at last met at Promontory, Utah. The historic last spike was driven. Two engines rolled in from each side and touched pilots; the shrieks of their whistles resounded the country over. A project that in 1856 had been declared as impracticable as a road to the North Pole was an actuality. The Union Pacific line had cost $39,000,000 to build; the Central Pacific, $140,000,000, and no more than forty years after the construction of the first American locomotive, the Union was bound securely together with hoops of steel.

CHAPTER XVII

Mr. Davenport Deceives His Wife

The nineteenth century was barely one quarter gone by before it became evident to the members of the Albany Institute that, in the field of electricity, the United States once again had an experimenter of the very first rank. He was Joseph Henry, in many particulars co-discoverer with Michael Faraday of the principles of electromagnetic induction. By 1827 Henry had delivered to the institute a paper "On Some Modifications of the Electro-Magnetic Apparatus." Soon he had so carefully watered and tended the dying seed of electricity planted in the United States by Benjamin Franklin that new shoots began flowering again in a dozen places.

In the mind of the village blacksmith at Brandon, Vermont, one of the first new shoots appeared. He was Thomas Davenport, and through the devious channels of communication he heard, sometime in the 1820's, of the experiments in magnetism with which Professor Henry was amazing the Albany Institute. Davenport's electrical interests, somewhat to the disadvantage of horses waiting to be shod, began then, and when he heard that one of the electromagnets which Professor Henry had built for the Penfield Ironworks, whereby it might sift iron ore more readily, was capable of lifting his anvil up by the roots, he could work no more until he saw it. In 1833 he said good-bye to his wife with a certain smirk of guilt and set off to Penfield. He must, he explained, purchase more iron, and only the Penfield Ironworks had just the kind of iron he needed. This, of course, was a fearful lie. The ironworks was fully twenty miles from Brandon—a long distance in those days. Thomas Davenport felt guilty, since he knew quite well that he could purchase iron much closer to home. But he had to see the magnet that could lift his anvil.

Now of course it could not. It was a good sinewy magnet and could catch up quite a number of pounds at one time. But the year in which Davenport visited Penfield was only two years after Michael Faraday had given his first demonstration of a dynamo. No dynamo had yet been built that could produce power on any useful scale; voltaic batteries were still the only steady source, and Professor Henry's magnet acquired its lifting power from them. These batteries were large and evil and spilled when you looked at them. The number of cells that could be brought together was limited and so therefore was the attractive power of the magnets which they energized.

But even if they would not lift his anvil they were wonderful enough to Thomas Davenport, these magnets. "Like a flash," said he later, "the thought occurred to me that here was an available power which was within the reach of man." Thomas Davenport returned to Brandon with enough iron to allay the suspicions of his wife and enough ideas to keep the world of invention busy for a hundred years.

His forge fire burned low and the shaggy mares and geldings of Brandon could only thump their hooves impatiently in his shop while the village blacksmith so far forgot his trade as to work on the discovery of "the production of rotary motion by repeated changes of magnetic poles." He was building the world's first practical electric motor. He finished in 1834, hitched it to a voltaic battery, and tried it out. It worked poorly, with a great deal of grumbling and spluttering.

Thomas Davenport was no scientist; he was only an ingenious practical man, and he had to learn the lesson that poor insulation of the wires in his armatures was causing him a serious leakage of power. But at least the motor worked and Thomas Davenport applied for a patent on it.

His claims were fine and resounding but his motor was frail and weak. He burned to build another, but blacksmiths are not wealthy, and no native of Brandon would lend money to such a crackbrain. Yet another motor must be built and there must be new material for insulation. Silk was the material he wanted but where was he to get it? . . . When he had been married, in 1827, his wife Emily had worn a pretty silk wedding dress.

Thereupon a desecrating thought occurred to Thomas Davenport. Where was that dress? It was obviously of no further use to his wife. All occasion was past; besides, the woman was too plump. For the second time at the command of that jealous mistress, Science, Thomas Davenport deceived his wife. On a pretext he found where it lay—carefully folded away in lavender and tissue paper as it had been for eight years. He sought it out, and the desperate glint in his eye as he marched with it to his workshop told his wife that her tears would be quite useless. She wept, in a tentative, experimental sort of way. But a hundred years ago a husband was master in his house. While his wife rocked to and fro in romantic grief, this unfeeling man cut the wedding dress into thin strips with sharp scissors and with them callously wrapped every foot of wire that went into his motor.

How long afterward Thomas Davenport was forced to sit in uneasy silence and meet over the rim of his coffee cup the reproachful gaze of his wife, history cannot tell. But the silk dress that was sacrificed in 1835 marked the coming of a time when ladies the land over might clean their rugs, sew their clothes, wash their dishes, beat their eggs, and dry their fair hair through the agency of "repeated changes of magnetic poles" upon which Thomas Davenport, his insulation now highly satisfactory, received a patent in 1837.

From then on the Brandon horses, in despair, took their trade to less imaginative blacksmiths. Thomas Davenport spent the rest of his life in building and experimenting with motors. In 1836 there was not a mile of steam railroad in the state of Vermont, but Davenport set to work and built miniature motors and tracks to show that if the state ever did have a railroad it could be operated by electricity as well as by steam. His car would travel on a circular track twenty-four inches in diameter with batteries in the center for power and the rails for the return circuit. Encouraged by his model's success, Davenport went so far as to suggest that a motor might be built to develop a full horsepower. At this, even Professor Henry stood aghast. He was very kind but he earnestly counseled caution. This idea, certainly, was a bit too fantastic.

Money was still a great problem to Thomas Davenport, but now at least his townsmen believed in him. In 1837 he organized the "Electro-Magnetic Association"—the first electrical stock company in the world. By 1839 he was even hoping to apply electricity to printing presses—just a quarter-century after the year when artificial power of any sort had first been applied to printing. The London *Times* had hitched its press to a steam engine on November 28, 1814.

But all this time the electrical experts of the day were placing the cart before the horse. No one seemed to realize that the discoveries of Faraday and Henry meant that power could some day be generated by dynamos on an even grander scale than motors could use it. Batteries remained in consequence the only source of power. The more ambitious engineers became, the more they became dissatisfied with the invention of Alessandro Volta which, in the year that began the century, had been celebrated as the most wonderful scientific discovery that the world had seen.

Even so, motors and rails, combined, were now producing a ferment. In 1847, ten years after Thomas Davenport's patent, Moses G. Farmer built an experimental electric car driven by wet batteries and capable of carrying two passengers—twice the number of anyone's previous ambitions. And in 1851 Professor C. G. Page, then successor to the great Professor Henry as secretary of the Smithsonian Institution, built a car to run on tracks from Washington to Bladensburg, Maryland. There were 100 slopping, green-crusted, acid cells in the batteries which drove Professor Page's motors but, alas for nineteenth-century roadbeds, the fearful speed of 19 miles per hour which this power developed set the little car dancing and bobbing so fearfully that after a short distance almost all of the 100 jars overturned and cracked, their acidlike blood draining hopelessly away. That ended a costly and visionary experiment. One George Green, a machinist in Kalamazoo, Michigan, tried the idea again in 1856. With a small battery car he could carry one passenger for 200 feet. But since, in that year, a steam locomotive could carry hundreds of people for hundreds of miles, there seemed no danger that the electric motor would ever compete with it. Curious, thought people, how learned professors and hard-headed machinists could make dunces of themselves at times.

None the less, the idea of electric traction would not down. Faraday's dynamo was slowly being made into a machine that would generate a useful current. By 1870 the Belgian, Gramme, had created a practical machine. Before the decade had ended, Thomas Alva Edison in the United States and Dr. Werner Siemens in Germany had once and for all brought it out of the experimental class and set it to work.

In 1879, at the great Berlin Exposition, Siemens astonished

the world. Near the river Spree there had been laid a track a third of a mile long. There were three rails, and the third lay between the wheel tracks. It was charged not by batteries but by a dynamo. For this, the first successful street railway system in the world, there were three cars. They could carry 20 passengers each at eight miles per hour. The incredulous world woke with a shock to realize that electricity was not solely a scientist's abstraction after all. It was a realization redoubled when, on May 12, 1881, the first permanent electric railway line was put into operation between Berlin and Lichterfelde, a distance of 1½ miles.

Now the electrical ferment was working everywhere—nowhere more earnestly than in Menlo Park, New Jersey. Thomas Alva Edison had created a successful incandescent lamp; his thoughts now were turning to other uses for electricity and better means of generating it. In the spring of 1880 he built a small line of 3½-foot gauge. It was shaped like a U and was a third of a mile long. Edison gave up the third-rail system of Siemens; he made his two rails a complete return circuit, furnishing them with current from two 110-volt dynamos previously used in his lamp experiments and of a type known among the Edison ranks as "Long Waisted Mary Anns." Each would produce 12 horsepower.

The locomotive which Edison designed for this railway looked rather like a track laborer's handcar. Its basis was a four-wheeled iron truck which carried a platform, upon which another Mary Ann dynamo was used as a motor—laid on its side so that the motor projected at the front. At first the motor was connected to the wheels by friction pulleys; later Edison replaced these with belting connections.

On May 13, 1880, the Edison locomotive growled over its humpy track for the first time, and the papers rang with the news. It could haul 26 passengers at a maximum speed of 30 miles per hour. Furthermore, here were 12 horses com-

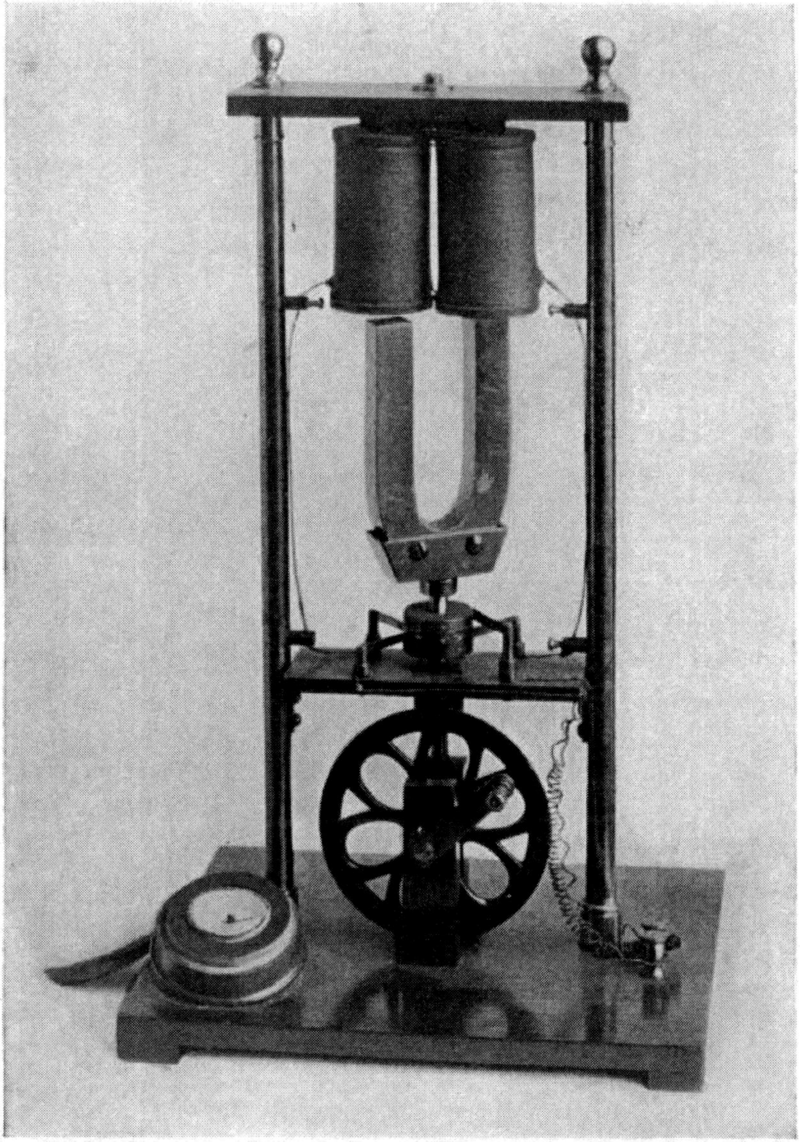

Smithsonian Institution

PIXII'S MAGNETO

ONE of the first dynamo-electric machines to be produced following Faraday's great discovery

Smithsonian Institution

ELECTRIC CAR

DESIGNED by Moses G. Farmer in 1847, it made early demonstrations of the traction possibilities of electricity

Smithsonian Institution

EARLY STREET CAR

LEO DAFT's electric engine and car, an early American fore-runner of modern street railway systems

pressed into a smaller space than ever, doing their work without the excretion of any smoke, ashes, or other nuisances —this only four decades after Joseph Henry had wrung his hands in high-minded scientific distress at the thought of a motor as gross as one horsepower. Things were coming on.

Particularly they came on in the newspapers. In June, 1880, the New York *Daily Graphic* published a sketch of a 100-horsepower locomotive which Edison would soon, so it was said, build to run on Pennsylvania Railroad tracks between those two centers of culture and the arts, Perth Amboy and Rahway. But the Pennsylvania Railroad thought otherwise. Edison had in mind as large a locomotive as 300-horsepower and had actually projected plans for one at his works on Goerck Street, New York. But when President Frank Thomson of the Pennsylvania visited Menlo Park and was shown the little railroad, now a trifle over a mile long, he swiftly decided that steam could never be supplanted. And for his time, he was quite right. It was to be fifteen years more before a steam road was first to use an electric locomotive.

Yet even in 1881 Henry Villard, president of the Northern Pacific, was certain that the problem could be solved. He advanced some $35,000 to Edison for experiments that would produce an electric locomotive with an economy better than steam and develop a speed of 60 miles per hour. The prospectus was impossible for the year, and Edison gradually drifted into electrical experiments other than traction. He did develop a locomotive that would travel 40 miles per hour but was otherwise none too satisfactory and had a bad habit of jumping into ditches.

In the meanwhile, another electrical experimenter had been hard at work. Charles J. Van Depoele had been born in Belgium in 1846. He was brought up in the mechanical way he should go by his father, a master mechanic of railway shops in East Flanders. Railroading appealed to the boy, but electricity did far more. By the time he was fifteen, he had succeeded in operating a crude electric light with the current of forty cells. But his father was impatient with this fooling, and by way of showing young Charles that he was not taking a serious interest in the work of the world he apprenticed him to a Paris cabinet maker who carved altars and statuary as his business. Charles retaliated in the only way he knew. He ran away from home by the device of visiting an aunt in Antwerp and never coming back. This was in the year 1868, when he was twenty-two.

He turned up next, this Flemish wood carver, in Detroit, Michigan, where, in the days before the automobile, the furniture trade was one of the great businesses of the city. He began a small factory and before long, such was his enterprise, he was employing two hundred men. At last, in 1877, he took a sweet and noble revenge on his father. Full of prosperity, he brought his parents to America and thereupon turned over to his father, the former master mechanic, the business of carving church furniture while he, the son, set out once again along the course he loved. He began spending the profits of the furniture business in experiments with electricity.

To follow up his earlier studies, he built a dynamo and a large arc lamp. One foggy night he tried it out with so much success that a nervous observer seeing its glow through the mists promptly turned in a fire alarm. This, thought Van Depoele, must be good. He built more and larger arc lamps, and in 1878 he made himself history's first authority in lighting for dramatic effect. Forepaugh's Circus came to Chicago; Van Depoele lit up the big top with one of his tremendous arcs. The sensation was equally tremendous in an audience that heretofore had had to take its drama flavored only with kerosene.

But the one invention for which Van Depoele will always be remembered is the system, or the ancestor of it, by which most power for electric street cars is carried and collected to-day. Previous to him, John C. Henry, a young telegraph operator, suggested that the dangers of exposed third rails or insulated tracks be done away with by erecting an overhead wire system. His idea was to set a small inverted carriage of four wheels on each of two overhead wires and attach it to the car by a flexible cable. This he called a "troller" because the car trolled it along like an upside-down fishpole in an upside-down lake. Two things happened to young Mr. Henry's invention. The common speech changed its name to "trolley," which seemed somehow simpler, and Van Depoele did away with its unnecessary complexity by substituting for it the simple little under-running wheel held against the wire by heavy springs attached to a stout iron rod—the mechanism which Oliver Wendell Holmes called "The Witch's Broomstick." Frantic for good public causes to perform, the newspapers of the day damned the inoffensive overhead wire as deadly and comic journals published dialogues like this:

Nervous Old Maid: "Officer, if I put my foot on the rail will I get a shock?"

Officer: "No, ma'am, not unless you put your other foot on the wire."

In spite of all, the troller prevailed. The force of electricity was now a servant of man. It would light the dark. It would carry him where he wished to go on its invisible river of power, and his craft was an odd, bobbing cork of a vehicle, the like of which is to be found now in only one place on the face of the earth: the comic pages of newspapers wherein Fontaine Fox, Jr., draws his inimitable pictures of the Toonerville Trolley.

CHAPTER XVIII

Mr. Edison's Jumbo

In 1820 the French scientist Dominique François Jean Arago magnetized a bar of soft iron by surrounding it with charged wires. In 1822 Michael Faraday wrote in his notebook the tremendous prospectus, "convert magnetism into electricity." Nine years later he thrust a magnet through a coil of wire and did it. He followed this by the construction of the first crude machine called a "dynamo" and the way was clear for the conversion of mechanical into electrical energy.

Men of science were now in the peculiar position of having invented a great, new machine without having any very clear ideas of what good it might do them. Even for the current from the voltaic pile there was little practical use, save Sir Humphry Davy's arc light—and the early dynamo was not even good for that. The dynamo, more than thirty years after its invention, belonged still to the great army of the unemployed.

But the Frenchman Pixii lost no time in following the road which Michael Faraday had pointed out. In 1831 he invented the magneto—a coil of wire spun around in the field of a permanent magnet. This crude device was little more than a curiosity, but Pixii improved it remarkably by two innovations. The field of a permanent magnet is weak; that of an electromagnet has much greater power to produce lines of force. Thus it occurred to Pixii that if he sidetracked a little of the current which his magneto produced, and ran this extra current through coils surrounding his permanent magnet, he would help matters, since the electromagnet thus devised would strengthen the field and therefore strengthen the current emanating from the machine. Having done that, he likewise invented the commutator—a ring composed of alternate metallic and insulating segments so that the current drawn off from Pixii's whirling coil of wire would no longer, in the nature of things, fluctuate from a positive value through zero to a negative value (and so on *ad infinitum*) but would have a direct, if slightly varying, flow in one direction. The normal current in the armature of a dynamo is always what is called "alternating"; that from a voltaic pile is always "direct," and at first it seemed that Pixii's invention had removed an enormous disability from Michael Faraday's dynamo. It was to be still a good many years before discovery came that this was not so.

Yet the dynamo, its feeble muscles rendered all the more flabby from lack of exercise, continued to languish. Then in 1870 the Belgian, Z. T. Gramme, greatly improved matters by developing what was known as the ring-wound armature. The result was the first practical generator to yield a satisfactory direct current. Around a soft iron ring Gramme wrapped a number of separate insulated coils, the ends of which he brought out to a commutator. When this ring was revolved in the field of a magnet, a current was induced in the separate coils, and as the commutator segments, one by one, passed the point where the brushes scraped against them, each separate little coil of wire shot in its small portion of electricity to swell the total.

Now, having broken down this one great barrier deliberately, Gramme blundered accidentally into another, and, without at all intending to, knocked that one flat also. In 1873 the great Vienna Industrial Exposition was about to open and Gramme's generators were to occupy a place of honor. Side by side there stood two dynamos, one already being driven by a steam engine—the other with its drive shaft not yet connected. A workman, not knowing or caring particularly what he was doing, dreamily connected the wires from the moving dynamo to the quiet one, whereat, to his terrified surprise, the quiet dynamo, still with no connection between itself and its steam engine, began to revolve as if the devil had possession of it. In this emergency Gramme was sent for, and after watching this queer spectacle for a moment the truth burst on him—a dynamo and a motor were not, as had always previously been thought, two entirely separate machines but were really one and the same. The moving dynamo was driving the other one as a motor. No dynamo had ever made a motor revolve before, but now it became swiftly apparent that Gramme's new direct-current machine would do all the things that a battery could do, and might, for that matter, do them a great deal better.

But of all the things that the Gramme dynamo set in motion, none was more useful than the mind of Thomas Alva Edison—who, in the year of Gramme's Vienna discovery, was a young man of about twenty-five. He had, it is true, sold newspapers in his youth and had been a candy butcher on the trains; notwithstanding, he was the son of well-to-do people who belonged among the socially élite of the little town of Port Huron, Michigan. His fertility of invention was already the talk of the country when he came upon the report of the Franklin Institute in Philadelphia, which sung the economies and advantages of the Gramme machine. It rendered high praise to Gramme's dynamo for the reason that "it utilizes in the arc [lamp] from 38 to 41 per cent of the motive power produced, after deduction is made for friction and the resistance of the air." If, thought Edison, this machine is so good, why should not affairs be organized to develop it upon a commercial scale? Let us give people the opportunity of buying something new—

electric current, continuously supplied in their homes. Edison had excellent reason for thinking so; by now his incandescent lamp was a success and all he needed was some system whereby current to light it might be generated and sold to the world at large.

With the combined enthusiasms of inventor and business man, he set to work. He had now the advantage of something which was making even the Gramme dynamo out of date—the drum armature first developed in 1872 by Von Hefner Alteneck. Edison drove a dynamo armature at varying speeds with different types of windings, until he had developed a machine which, he claimed, showed a 90 per cent efficiency. By 1879 he had one which satisfied him.

In 1881 there was to be held in Paris the first great electrical exposition, and Edison, who was always a genius at publicity as well as at science, determined that he would design and build a dynamo and ship it to Paris for this international scientific fête. His time was short. He set feverishly to work and soon had completed a machine which, with its engine, weighed 27 tons, and of which the armature alone weighed 6 tons. It could light, when it felt in the mood, no less than 1,200 incandescent lamps at once. It was the "Edison Jumbo," the eighth wonder of the scientific world.

Yet all was not well with it. On its first test its voltage proved too low, and Edison was forced to make hasty alterations in the two electromagnets of the field, whose enormous size gave his machine its name. There was no reason whatever why four, six, eight, or ten magnets could not now be applied to the dynamo, but such was traditionalism, even in a mind as gifted as Edison's, that he and everyone else clung to the bipolar idea—only because that was the form which an ordinary horseshoe magnet had always taken.

Even the strengthening of the field coils did not end Edison's troubles. When next the machine was tested the crankshaft of the engine broke and hurled itself across the room like a high explosive shell. By the time this damage was repaired, only three days remained before the Jumbo machine must set sail. Another test had to be run and it was quite likely that this would take every remaining moment.

There now enters into the history of electrical machinery the name of Tammany Hall, that unique New York political institution. To a Tammany leader Edison made an appeal. Let the route from his shop to the ship's pier be lined with policemen to keep traffic clear—so that the horse-drawn trucks which were to deliver the dynamo to the pier might make their journey over the rough cobblestones unimpeded by the fiacres and barouches of the day.

> About four hours before the steamer had to get it, the machine was shut down after the test [said Edison], and a schedule was made out in advance of what each man had to do. Sixty men were put on top of the dynamo to get it ready, and each man had written orders as to what he was to perform. We got it all taken apart and put on trucks and started off. They drove the horses with a fire bell in front of them to the French pier, the policemen lining the streets. Fifty men were ready to help the stevedores get it on the steamer—and we were one hour ahead of time.

Now, while this first Jumbo distinguished itself before incredulous crowds in Paris, Edison settled down on Goerck Street, north of Grand Street, New York, to manufacture dynamos in the old Aetna Iron Works. Here, the "Edison Machine Works" turned out two more machines. These were shipped to London and installed there in 1881 with the idea of creating the first central station in the world. It was built upon Crown land and on January 12, 1882, it was set in operation to light 3,000 incandescent lamps from a power plant located on the gloomy Holborn Viaduct in London City. Here, perhaps, might have been the great beginning of the central electrical power system in Europe, save that the English, exhibiting a strange passion for legislative restrictions, passed the Electric Lighting Act of 1882, which throttled it out of existence in favor of the continuance of gas lighting—itself an interloper no more than sixty years old.

But the idea of generating, transmitting, and selling power was not now to be denied. Even before Edison's Jumbo was a reality, a small hydroelectric plant had been set up in Meisbach, Bavaria, and had transmitted a feeble amount of power over iron wires leading to the great city of Munich. Then, on August 15, 1882, the first faltering hydroelectric plant in the United States came into being at Appleton, Wisconsin, where a small dynamo, housed in a wooden shed and just strong enough to light 280 lights of ten candle power each, began its feeble and uncertain revolutions—which were to continue until the year 1899.

Meanwhile Edison continued his hot pursuit of the central station idea. The "Edison Electric Illuminating Company of New York" was organized on December 17, 1880, with the original idea that sufficient Jumbo dynamos would be built to supply the entire area from Wall Street to Canal Street with electric light. But this was too ambitious, and the territory which Edison finally staked out was bounded by Wall Street, Nassau Street, Spruce Street, Ferry Street, Peck Slip, and the East River—approximately a square mile. Through this district he sent canvassers upon a house-to-house investigation to discover how many gas jets every building had, how many hours, upon the average, they were used, and the cost of the gas they burned. Every hour of the night, up to three in the morning, he sent men through this district noting the number of gas jets still alight.

Slowly customers volunteered for the proposed new service and the Edison Company now began digging up the streets to run its wires through underground conduits. Meanwhile, at 257 Pearl Street, Edison was at work building and installing six Jumbos. The first three, of which the largest was 125 horsepower, were installed in May, 1882, and were started on July 5th. They ran at 220 volts and on July 8th they proved their mettle by lighting 1,000 test lamps on the upper floor of the station.

At three p.m., on Monday, September 4, 1882, the first commercial central station began to supply customers with electric power. There were some fifty-nine of these customers who had been beguiled into using this curious new service through the promise of receiving three months of current at no charge. By the end of four months the Edison Company was lighting 225 houses in the district with a total of 5,000 lamps.

One early customer astounded the entire city by placing an order for an installation of 250 lamps. What for? thought Edison, himself amazed. The sad truth then transpired that his great new idea was already being made accessory to the Powers of Evil. This customer was a distiller of cheap whiskey, even then affectionately known along the Bowery as "rot gut." He longed for electricity out of his conviction that if he thrust a cluster of bulbs into a barrel of his whiskey it would accelerate the aging process and remove the "shudder." There was a momentary clash of interests in Mr. Edison's bosom, the instinct of the prohibitionist grappling mightily with the instinct of the business man anxious to sell his product. It was a brief encounter. Thereafter a fraction of the Jumbo's power was applied to the immoral purpose of making whiskey more palatable.

The Pearl Street station had gone into service with nothing more than minor difficulties, but still the Jumbos were each used separately. Edison had not yet dared to see what would happen if he tried the scheme (which had great practical advantages) of operating two of them in parallel on the same line. Finally on a Sunday, when business houses were closed and there was very little demand for current, he determined to make the experiment, later confessing that his heart was in his mouth as he did it. The first Jumbo started uneventfully.

> Then [said Edison] we started another engine and threw them into parallel. Of all the circuses since Adam was born, we had the worst then! One engine would stop, and the other would run up to a thousand revolutions, and then they would see-saw. The trouble was with the governors. When the circus commenced, the gang that was standing around ran out precipitately, and I guess some of them kept running for a block or two. I grabbed the throttle of one engine, and E. H. Johnson, who was the only one present to keep his wits, caught hold of the other, and we shut them off.

Within this brief interval the thunders and lightnings of Mr. Edison's two antagonistic elephants had all but ruined the country's first electric lighting system. The trouble, as Edison shortly discovered, was not solely with the governors, but lay in the fact that unless there were some positive way of keeping both Jumbos running at exactly the same speed the faster one would begin to drive the slower one as a motor, and the slower one, thus becoming the faster, would then return the compliment. It was some time before the experiment was repeated.

In the second year of the station's service the world of finance officially crowned it; the New York Stock Exchange, hitherto gas-lit, was put on the new circuit. With increasing demands it soon became evident that the generation of electric power commercially was no picnic. A dark day, producing a sudden demand for light, would overload the system so badly that the entire power house itself would slowly approach incandescence, as the overloaded armatures reached a dangerous heat. Furthermore, transmission difficulties appeared in unexpected ways and places. Soon after the Pearl Street station had been started, a policeman one day rushed into the generating room with the demand that someone be sent immediately to the corner of Ann and Nassau streets. The neighborhood had gone crazy, and electricity was suspected.

Edison went, and found the troubled area surrounded by an immense crowd. The cellar of a house upon the corner extended under the street, and in that cellar a junction box had gone wrong. The moist earth above it gratefully took up an electric charge, and when the horse of a rag man had sauntered sedately down the street, the spirit of Pegasus awoke in his withers as, through his iron-shod hooves, he was suddenly given a strong jolt of electricity. This had happened several times with other horses, and the watching crowd had grown to such proportions that Edison had no alternative but to deprive the city of New York of its electric light until the junction box might be repaired. Next day a spectator of this scene called upon Edison to request that the defective junction box be installed in his own establishment. He was a dealer in horses, he said, and saw a fortune in it. Sufficient electricity will, for a few moments, make any nag brother to a thoroughbred.

Customers likewise made the operation of the first station a trying business. The meters of the day were chemical affairs which measured the amount of current supplied by the amount of metal which the current would deposit upon a plate while lights were lit. No less formidable a customer than J. Pierpont Morgan, the elder, summoned Edison to his office and made thunderous demands on him to know why he was being overcharged. Morgan, as a financier, was quite certain that the meter was overindicating. Edison, as its inventor, was quite certain that it was not. The argument was not settled until every lamp in the Morgan establishment was provided with a tag upon which whoever turned the light on or off would note the time. For the first two months this tally, which the richest banker in the world scrupulously maintained, showed a discrepancy. Morgan clamped a cigar closer between his teeth and demanded a rebate; Edison squared his jaw and refused it. Triumph finally perched upon the Edison banners when he was able to show that night cleaners, unaware of the controversy, had kept lights burning at night without notations upon the tags. Morgan, for once in his life worsted in a matter of finance, withdrew his claims. And the first central station in the world thereafter continued a comparatively unexciting existence until, on January 2, 1890, it was partially destroyed by fire. All the Jumbos, save the famous Number 9, now a relic of the present New York Edison Company, were reduced to shapeless twists of metal.

But Edison now had other greater matters on his mind. On March 3, 1883, an electric light station was opened to supply power in Milan, Italy—the first permanent European installation. It grew until ten Jumbos were supplying power to a total of some 10,000 lamps, and was operated continuously until February 9, 1900. Meanwhile, in Brockton, Massachusetts, Edison opened a station on October 1, 1883, and so quickly were the marvels of the modern world now following that when a new theater was opened there in 1884 it was never piped for gas. Here also electric current was supplied to the first fire house, so that the ringing of an alarm automatically lit every light in the place and released the horses for their charge. What was the world coming to?

CHAPTER XIX

The Emancipated Mules

The electric car seemed, to the urban world of the 1880's, the finest flowering of mankind's inventive genius. Curiously, although the principle of electric traction is only now beginning to come into its glory (the electrification project of the Pennsylvania Railroad is the most spectacular to date), the street car has already, aged fifty years, begun to die. Enthusiastic gentlemen with sideburns and wonderstruck ladies in bustles and bonnets and carrying small pink parasols could scarcely have credited this in the year 1883 when the Chicago Railway Exposition burst upon a city still in convalescence from the fire which had leveled it only twelve years before.

The history of what they saw had begun a few years earlier. It had originated in the mind of Stephen D. Field, nephew of Cyrus W. Field, who laid the first Atlantic cable. To the younger Field it seemed that electric motors could now efficiently replace the noisy and expensive cables which were forever hauling cars up and down the hills of his native San Francisco. This was a somewhat over-optimistic thought for the year 1877; many a city has not even yet completely forsaken its cable cars. But Field was as brave and long-sighted as his uncle; to prove his idea he ordered a dynamo from the Siemens firm in Berlin. It was built after many weary months and at a great expense, then shipped to the seaboard and started upon its journey across the Atlantic in a sailing vessel. Whereupon the vessel was lost at sea.

Field was all but ruined by the disaster, yet still he ordered a second dynamo and saw it safely delivered. By 1883 the Electric Railway Company of America had been incorporated in New York with a capital of $2,000,000 to develop the patents and inventions of Field and of Thomas Alva Edison together.

It was the result of this which the ladies and gentlemen of 1883 saw in Chicago. Around the outer edge of the gallery in the main building of the Chicago Railway Exposition a track of three-foot gauge was constructed—running to that mystic length of one third of a mile which no early electrical engineer could escape. Between the gauge rails there was a third rail, dynamo-charged to supply nourishment to an electric locomotive named "The Judge." "The Judge" was ponderous for that day; he weighed 3 tons, was 12 feet long by 5 feet wide and could carry a trailer filled with passengers at 9 miles per hour. Beginning on June 5th and running until the Exposition closed on June 23d, "The Judge" was operated for 118 hours, traveled 446 miles, and carried the tremendous number of 26,805 passengers. So great a success was this exhibit that it was rebuilt for the Louisville Exposition in the same year. It was again triumphant.

Doomed by the new invention, horse-car lines had none the less been thrusting out their steel arms everywhere. Between 1850 and 1855 half a dozen lines had been built in the cities of the United States. In 1860, when this country had nearly forty lines, England achieved her first—American-built by George Francis Train at Birkenhead. One more decade, and by 1870 the United States could boast that its cities offered transportation facilities equaled by none other in the world, for no less than 120 companies operated gaudy horse-drawn cars over uneven rails sunk in the cobblestones. It was fifty years before that Michael Faraday, with a great discovery locked in his teeming mind, had watched the dancing horses of Astley's Circus. It was time that someone redeemed the promise of that night that the horses of the future might dance more and work less.

So far as the United States is concerned, the first step toward redemption began when Leo Daft (it is all too easy to see the use that scoffers would make of his surname), a clever English photographer, came to this country and became interested in electricity. Soon he was building the electric locomotives which, with touching sentiment, he named in succession, "Morse," "Volta," "Ampère," and "Benjamin Franklin." Then, in August, 1885, he began to equip the Hampden branch of the Baltimore Union Passenger Railway Company for electrical operation with a third-rail system. The contract specified satisfactory operation in all particulars before Daft was to receive a penny in payment. Quite naturally, engineers told him that only a fool would enter such an undertaking. Notwithstanding, Daft finished his work to the satisfaction of everyone, and the Hampden branch became what is probably the first regularly operated electrical street railway in America.

Encouraged by the Hampden success, Daft undertook a new project. He would, he said, demonstrate to the Metropolitan Railway that its elevated trains, which were now filling the Borough of Manhattan in New York with their rumblings, could be successfully electrified. In this year the elevated trains were drawn by "Forney" engines—curious steam locomotives with four wheels carrying engine, coal, and water upon one frame—the water in a large iron saddleback slung over the boiler. They weighed nine to ten tons and were built to scuttle around the sharp curves of the system.

Daft began his work by electrifying a two-mile section of the Ninth Avenue Elevated. There he used the "Benjamin Franklin," his latest locomotive, on several trips. The "Ben-

MODEL "JUMBO"

ONE of the grandfathers of the modern dynamo, as designed by Thomas Alva Edison

Smithsonian Institution

"JUMBO" AT WORK

ONE of Edison's early dynamos belt-connected to a steam engine

Smithsonian Institution

New York Edison Company

EARLY CENTRAL STATION

A MODEL of Edison's Pearl Street station which in 1882 supplied New York City with its first commercial electricity

New York Edison Company

DYNAMO ROOM

PRINT showing the interior of Edison's Pearl Street station

jamin Franklin" was nothing much more than a motor mounted upon a platform and connected to the axle by friction gears held in contact by the weight of the machine and regulated by an adjusting screw. The experiments were soon suspended—but even a machine with the name "Benjamin Franklin" had a quality of persistence. The experiments were resumed again in 1888; in that year the rebuilt "Franklin" for several weeks made trips running between steam trains on a section between Fourteenth Street and Fiftieth Street. It could reach a maximum of twenty-five miles per hour and once distinguished itself by pulling an eight-car train up a 2 per cent grade at seven miles per hour. Unfortunately, Daft was committed to low-voltage design, which at the time militated against high speeds, and these inconclusive but praiseworthy experiments soon came once again to an end. Thus upon the city streets, the hard-working bays and sorrels of the horse-car lines still waited for an outstanding deliverer. He turned up, of all unlikely places, in the United States Navy—Lieutenant Frank Julian Sprague.

Sprague was born in Milford, Connecticut, on July 25, 1857. In 1878 he graduated from the United States Naval Academy. But naval life provided, to his active mind, too slow and tortuous a path to success. Lieutenant Sprague found insufficient outlet for his abilities in routine maneuvers. What he was interested in was electricity.

Particularly he yearned to visit the Paris Electrical Exhibition of 1880–81. His request to be made an assistant to the navy officer detailed there met with a loud, round "No." His superiors suggested to him, however, that he could get himself ordered abroad and go to Paris at his own expense while there. He seized his opportunity, and during his passage made every attempt to install an electric bell system on board the ship. But the regrettably uncertain qualities of early electrical circuits made it quite likely that the captain upon summoning his navigating officer would instead be favored with a visit from the cook. This early experiment in communication was abandoned.

To add to Sprague's disappointment the exhibition had ended by the time he reached Paris. He was consoled only by being ordered, in the following year, to the Crystal Palace at Sydenham as secretary to the jury of awards at the Electrical Exposition held there. Part of his work was the testing of all electrical machinery to be installed.

Sprague was now in clover. Particularly he was impressed by the exhibit of Thomas Alva Edison; a personal and scientific friendship began to spring up between him and one of Edison's chief assistants, Edward H. Johnson. What impressed him even more, however, was the system of transportation that cursed some parts of London. The mighty flow of traffic in the City had by that time proved too much to be borne upon the surface of the earth, and early undergrounds were already in operation. They were steam-powered, and the impenetrable murk, stifling fumes, and fearful glare from open fire boxes made them the closest approaches to hell that mankind has ever voluntarily created. Sprague was forced to use them, and as he choked one day in their deathly, sulphurous air, he could not help but think what a boon for travelers in cities the electric motor might become. When he returned to the United States in 1883, it was with a mind actively engaged upon the idea of electric traction.

The navy broke into this daydream by ordering him back to his ship, but a disinterest in battle maneuvers had now risen to a sharp distaste. When Johnson offered him employment in Mr. Edison's company he joyfully accepted and reported for work on May 24, 1883—the historic day that Colonel Roebling's Brooklyn Bridge was first opened to traffic. Johnson had promised Sprague the salary of $2,500 a year. Edison grumbled a little at this extravagance but paid it. Once again, however, Sprague found himself doing things he did not enjoy. He was assigned to work on lighting developments, despite his belief that the future of electricity lay in traction fully as much as in illumination.

Sprague brought mathematics into a domain which Mr. Edison was ruling by more empirical methods. The Wizard of Menlo Park raised his brows at this but, none the less, put Sprague in charge of the new three-wire station at Brockton, Massachusetts—where the first theater was built for lighting by electricity only, disdainful of Welsbach burners. By correspondence there now began a wrangle between employer and employee, Sprague pleading that there was more money in selling power than in selling light. Edison retorted by suggesting that Sprague might be happier if he resigned. Sprague thought so, too. Working for Mr. Edison was turning out a little too much like being in the navy; a man's mind was not his own. Thus, in 1884, was formed the Sprague Electric Railway & Motor Company, with Edward H. Johnson furnishing the necessary capital.

In this year Sprague made his first great improvement in traction. In early electric cars the one motor had been mounted squarely in the middle of the chassis, where it produced a huge tumor-like bulge. It was Sprague who hit upon the now universal "wheelbarrow" method of suspension.

Then came a magnificent opportunity—a chance like the one Leo Daft had had to experiment with the possible electrification of the New York elevateds. In an old sugar refinery on East Twenty-fourth Street there were a number of Edison lighting generators. Here Edison and Field had conducted some experiments upon a track laid in a narrow canyon between two walls of the refinery. Sprague took over this equipment and made his first tests by mounting two motors upon an elevated railway truck under a flat-car and running it back and forth on its meager two hundred feet of track, drawing current by a small wheel in contact with a middle third rail.

Jay Gould was one of the great financiers of his day—a day when business ethics were not what they are now. His fortune had been built upon stock manipulations of railroads. Accordingly he was regarded as an expert in transportation. In his capacity as one of the principal owners of the Manhattan Elevated, he was invited to ride upon Sprague's test car. The ride had barely begun when a main fuse (for which, perhaps, the weight of Mr. Gould's reputation was too heavy) blew out with a flash. This was part of the day's work for Sprague, but it was the equivalent of Gabriel's last trump to Jay Gould, who was physically restrained (it was not hard, for he was a small man) from jumping to what he considered a point of greater safety. The fuse was swiftly replaced, but there was no possible repair to be made to Jay Gould's nerves. Thereafter he had little interest in Sprague's ideas of electrification. Although the scene of these experiments was transferred in May, 1886, to the Thirty-fourth Street spur of the elevated which ran between the East River and Third Avenue, no director of the railroad manifested any interest in applying Sprague's experiments to motive power. The great Cyrus W. Field observed them; so, from the eminence of the British peerage, did the Duke of Sutherland. But Sprague was not interested in a social success for his electrified car; he wanted a financial one, which seemed not to be forthcoming. He abandoned the idea of electrifying the elevated and hunted for more sympathetic fields.

Nevertheless the electrification idea was spreading. Boston, New York, and Philadelphia were toying with cars driven by storage batteries. One day in August, 1877, the New York *Sun* published a slightly bewildered but highly significant paragraph under the title of "Electricity on Wheels." This was it:

> They tried an electric car on Fourth Avenue yesterday. It created an amount of surprise and consternation from Thirty-second Street to 117th Street that was something like that caused by the first steamboat on the Hudson. Small boys yelled "dynamite" and "rats" and made similar appreciative remarks until they were hoarse. Newly appointed policemen debated arresting it, but went no further. The car horses which were met on the other track kicked without exception, as was natural, over an invention which threatens to relegate them to a sausage factory.

This was not of Sprague's doing. His glorious chance seemed to have vanished in the flare of the blown fuse which frightened Jay Gould. Then in May, 1887, a New York syndicate summoned him with a startling suddenness for the construction of a complete street railway system, in Richmond, Virginia.

Of the "Richmond Union Passenger Railway" there existed nothing. Sprague's contract called for the construction of a steam-generating plant of 375 horsepower, erecting overhead wires and feeders, and equipping forty cars with two 7½ horsepower motors each. There were no tracks. There was barely a route. Twelve miles of track with twenty-nine curves would, so Sprague was promised, be laid by another contractor. For his share of the work Sprague was to receive $110,000—if the financiers for whom he was doing the work later judged that it was satisfactory. Furthermore, the entire project was to be completed in ninety days.

Here was a stupendous job. Up to this time there were probably not more than eighty motors in use for electric traction the world over. To simplify everything Sprague then came down with typhoid fever, and for nine of the thirteen working weeks he lay muttering feebly to himself in bed and directing the work of his associates between spoon-feeding of gruel and other indignities.

Yet the job was done. On February 2, 1888, the Richmond Union Passenger Railway began operation. It had been informally inaugurated the day before when some psychological genius set the cars going to carry all the children of the city free of charge. Prejudice, and there was much of it, was partially broken down.

So, even as it began, was the trolley line. The rails were badly joined and laid without foundations. The power house was quartered in two rough sheds for its unprecedented work of generating 450 volts. Yet Sprague's motors and generators performed, even at the task of carrying their cars up a 10 per cent grade. Sprague put them to the test with fear and trembling on a dark night to be as free as possible from spectators. Although one motor ruptured itself in the task, the car did reach the summit. For his pains Sprague eventually received $80,000 of his promised $110,000—having spent $150,000 on the job.

Then, to add one more complication, the telephone company discovered that the transmission of traction power seriously disturbed its own overhead lines through the quality of "sympathetic induction." It promptly went to court to have the nuisance abated. It was a battle fought in twenty-seven states but settled, as common sense might earlier have suggested, in the laboratory, not the jury room.

Sprague was famous now. New Orleans demanded that he be summoned to emancipate the mule from street railways as Lincoln had emancipated the slaves. Sprague was not only a practical constructor of the first satisfactory streetcar system in the United States; he was the inventor of such essential adjuncts to operation as the series-parallel coupling of motors for operation over the proper range of speeds. And his greatest contribution was still to be made.

This was the so-called "multiple unit control," whereby, to-day, a motorman in one end of a train can control the operation of a number of motored cars with the turn of one controller. Sprague had hit upon the idea during the time when a temporary eclipse from the traction field had turned his thoughts to elevators. It was an idea of sufficient force to bring him back to his first love. He saw that through it power could always be kept proportional to load and that the entire weight of the train could be utilized for the driving wheels. He plunged into a mass of complicated calculations and diagramming so effective that when the Edison-General Electric Company completed the manufacture of the equipment at Schenectady (it was to be used for the South Side Elevated Railway in Chicago) Sprague's ten-year-old son, on July 26, 1897, by one turn of his hand was able to operate a six-car train upon the test track. In November of that year five-car trains went into operation in Chicago. Three months later the steam locomotives of the city were entirely abandoned and 120 cars went into multiple-unit operation.

It had been a great victory and Sprague had won it almost single-handed. So completely was the mule to be emancipated that a day was coming when the subways and elevateds of Manhattan Island, alone, were to collect some four million fares every twenty-four hours and carry, in the course of a year, twice as many passengers as the combined steam railways of all the United States.

CHAPTER XX

The Singing Wires

THE practical usefulness of electrical energy was now a certainty. But out of this certainty three new uncertainties now arose. In the first place, what was electricity's great use to be? Second, which was the better form in which to generate it, direct or alternating current? Third, how might it be efficiently transmitted?

The first of these questions had begun to be answered by Edison and Sprague. But there remained the second question: in what form was electricity to be transmitted for use? Because the wires on a dynamo armature are alternately cutting lines of force from north and south magnetic poles the normal current that is developed in an armature travels first in one direction and then in the other; i.e., it alternates. Early engineers, thinking this an almost insuperable objection to electricity, had spent great time and patience in the development of the commutator—a device whereby these alternations were reversed whenever they approached the zero line, and a rapidly pulsing direct current artificially created out of the normal, alternating one.

After a while it occurred to a few quizzical souls to wonder whether the alternating current might not have advantages along with its defects. The result was a long and bitter scientific quarrel, with Edison in the United States and the great Lord Kelvin in England leading the fight for direct current in opposition to men like Elihu Thomson and Nikola Tesla in the United States who believed in alternating current. Only within comparatively recent years did electrical engineers definitely realize that, in developing a dynamo to generate only direct current, they had been in the mistaken mood of a banqueter throwing away the most succulent meat of the lobster and resolutely making a banquet off the shell. Yet there were good reasons why this was not at first apparent. Early motors would not run on alternating current. The important work of electroplating could not (and never will) be performed by alternating current. It will not charge a storage battery.

Yet alternating current did have enormous advantages even before the days of appliances which will run on nothing else. In many ways it was Elihu Thomson, son of an English machinist who had migrated to the United States, who made them recognized. It was Thomson who raised a humble voice against Lord Kelvin in the fight for alternating-current machinery. It was he also who, in 1879, developed the so-called "three coil" dynamo for lighting arc lamps—a direct-current machine but one which shortly thereafter led the way toward alternating-current dynamos and motors—and who followed this splendid achievement with the so-called "repulsion motor," a cousin to the single-phase induction motor. Thomson's work not only made possible new scientific advantages in electricity; upon it was reared one of the three foundations of the enormous General Electric Company of to-day.

It was Nikola Tesla who advanced the battle upon still another front. Tesla, born in Smiljan (now a city of Jugoslavia) in 1857, was the son of a Greek-Catholic clergyman who had ambitions that his son should follow him in the church. To this, young Nikola respectfully objected. There was a serious impasse, interrupted only when cholera broke out in Serbia. The natives of the community in which Tesla lived burned large piles of odorous shrubbery to purify the air, but continued drinking copious drafts of the infected water which was spreading the epidemic. Along with almost everyone else, Tesla was stricken. For nine months he lay ill, several times at the point of death, but not too feeble to persuade his father that if he recovered he might study engineering instead. His father promised, Tesla recovered, and by 1877 was embarked upon his student career at Gratz, in the Austrian province of Styria—rising at three every morning and studying until eleven every night.

In the winter of 1878 Tesla's school laboratory was gratified to receive a new dynamo—a machine with a Gramme ring-wound armature and laminated permanent magnet that acted as the field. Tesla's learned professor one day demonstrated to the class the principle of self-excitation by winding a wire around the field and using a battery to drive the machine as a motor. Tesla noticed that the arcing, spluttering brushes in a great measure destroyed the efficiency of the machine. He thunderstruck his professor and the class by suggesting that such things as brushes ought to be eliminated.

This was heresy; the professor, like all challenged conservatives, had a pat answer. Any such suggestion was but a step removed, he said, from the demonstrable absurdities of perpetual motion. The class laughed heartily and Tesla, red with embarrassment, decided then and there that his idea was not silly and that some day, by the grace of God, he would prove it. He spent the rest of his term at Gratz in intense and fruitless effort.

In 1880 Tesla transferred his studies to the University of Prague. In 1881 he went to Budapest as a draftsman for—of all concerns—the Bell Telephone Company, which was then rigging the Dual Monarchy with its wires. He sat one day in a city park with a colleague, thinking nothing about electricity, but busily reciting portions of *Faust*. He reached a climax—and all of a sudden the sins of Mephistopheles vanished in a flash out of his mind. Into it there came instead a lucid diagram of a brushless motor.

This motor, of course, depended upon alternating current. A direct-current armature revolves because its own windings are perpetually trying to assume one position in the field and are being perpetually disappointed by the commutator. Now, thought Tesla, let the field coils be arranged so that the magnetic forces themselves, due to the alternations of current, chase themselves around perpetually and, just as perpetually, drag the armature with them. A direct-current armature was like a horse being urged on by carrots in front of his nose. Tesla's armature was like a white rat eternally trying to get somewhere else by galloping madly inside a rotating cage.

For a year, Tesla went through an agonizing period with too little leisure from his duties (he was now chief electrician for the telephone company) to get his motor built. On a visit to Paris in 1882 he met the American, D. Cunningham, who was sufficiently interested in Tesla's ideas to offer to form a company for commercializing them. Unhappily, Tesla's duties whisked him away to Strassburg before anything could happen.

But there, finally, he got his first motor built. It was a crude affair, this machine of 1882, but it ran, and there were no brushes to it. In 1883 he built two more machines of improved design, and in 1884 he went again to Paris, seeking money for the exploitation of his idea. The French proved to be not a good race for the encouragement of such ideas, as the unhappy ghosts of Cugnot and Jouffroy could have told him. Tesla got no money and, getting none, resolved to come to the United States—the land, as he had heard, of opportunity. In America Tesla quickly made two important associations. With Thomas Alva Edison he sought to talk of his new motor in the academic shades of Coney Island. The intervention of a tramp made this particular attempt unsuccessful, but even later Edison continued to favor direct-current machinery.

Even more important was Tesla's association with that extraordinary man, George Westinghouse. It was Westinghouse who, in large measure, brought to the United States the solution to electricity's third riddle—the riddle of efficient transmission. Here was a difficult matter. Resistance in a wire brings about a reduction of electrical pressure, or voltage, for exactly the same reason that resistance inside a water pipe lowers the pressure with every foot that the water travels away from its pump. Ohm's law is the mathematical expression of this melancholy fact. This difficulty was particularly serious for electricity, which was expensive to produce. Moreover, a heavy drop in voltage meant that the machinery at the operating end of the wires would run with a greatly lessened efficiency or else would not run at all.

Happily the answer to these difficulties lay in an experiment which Michael Faraday had performed in the year 1831. During the course of his great researches on electromagnetic induction, he had wound two coils of wire upon a soft ring of iron. When he sent a current through one coil he produced a current (an "induced" current) in the other, despite the absence of any physical connection between the two.

But no practical electrical engineer had found much use for this observation until, in 1884, the Frenchman Faulard and the Englishman Gibbs built a toweringly important new piece of electrical machinery. They called it a "secondary generator" (known now as a "transformer"), an instrument which, although having nothing to do with the creation of electrical energy, can exert profound influences upon its characteristics.

The power in a simple electrical circuit is expressed in terms of watts—a multiplication of pressure (volts) by rate of flow (amperes). A transformer, which is nothing more than Michael Faraday's two coils of wire, especially arranged in the relationship of their windings, will always give out from one coil the energy put into the other, with only a small deduction for losses. Thus, by arranging the number of turns of wire in each coil to suit their purposes, Faulard and Gibbs found that they were able to take a current, let us say, of two volts and six amperes and transform it into a current of six volts and two amperes. The power remained the same but the characteristics, for the purposes of transmission, were very greatly improved. That is because the higher the voltage in a transmission line the more efficiently the current in that line will be transmitted. If, then, Faulard and Gibbs discovered, one transformer (a "step-up") were inserted in a transmission line at the generator end and another (a "step-down") inserted where the current was to be used, the line between these two transformers could be made much longer than ever possible before, without serious losses.

This discovery was important not only for itself but for the interest it aroused in the mind of George Westinghouse, who had seen the results. Westinghouse hurried back to the United States, and with a brilliant young engineer by the name of William Stanley, Jr., tackled the problem of efficient electrical transmission. In 1885 at Great Barrington, Massachusetts, they built an experimental installation and found that they were able to step-up a current to one thousand volts, send it efficiently over a transmission line two miles long (this doubled the previous record), and then step it down sufficiently to light three hundred ordinary incandescent lamps. Ground was broken now for the great transmission lines of the future, and Charles P. Steinmetz had a foundation upon which to rear his awesome theories of alternating-current generation and transmission.

As a capstone to his transmission work, Westinghouse put into operation the first alternating-current central station in the world. It began its work at Buffalo, New York, in 1886 and was the harbinger of even greater things to come. When the first electrical power was drawn from Niagara, the electricity generated by this tremendous fall of waters was produced by Tesla two-phase alternators built by the Westinghouse Company. It was a modest design, made in 1895, with units of but 5,000 horsepower; to be compared with the 37,500 horsepower which is commonplace for a single generator unit to-day.

The man who helped create the early masterpieces of power at Niagara was a boy in his teens when Tesla first thought of his induction motor. He was Benjamin Garver Lamme, born near Springfield, Ohio, in 1864. At sixteen Lamme had puzzled mightily over a description of the Edison dynamo published in *Harper's Weekly* for January 3, 1880. That was an important date not only for the youthful Lamme but for the equally youthful dynamo. Before another sixteen years had gone by, the greatest generators of Niagara were to be designed by this young man and were to usher into the history of power a new era of significance and importance.

Lamme was born with one addiction which was to prove of extraordinary importance in electrical machinery. He loved high speeds. As a boy he once created a mechanism out of a wheelbarrow, a grindstone, and a spool. He turned the wheelbarrow upside down and belted the grindstone to its axle. To the wheel of the barrow he belted a spool held upon a shaft running between wooden stakes driven into the ground. With this combination he could, by turning his grindstone wheel 150 revolutions per minute, make the spool revolve at a speed (in those days even unsuspected by the future inventors of the turbine, De Laval and Parsons) of 30,000 revolutions per minute—provided nothing slipped. Young Lamme had not considered slippage. Neither had he considered lubrication. Thus when first he sought to justify his mania for high speeds his spool caught fire from the friction of its wooden pivots. Now he tried a metal shaft, and ground away terrifically at his handle. The spool first growled, then hummed, then shrieked. Having protested all it could, it lost every vestige of its self-control and, as mysteriously as a billiard ball in the hands of a magician, vanished. Nor did the young engineer find it again or the slightest fragment of it. Some years later, when new mysteries were made manifest to him, he realized that it had burst of centrifugal force.

Having put away childish things, Lamme graduated from the Ohio State University in 1888. Thereupon, he went to work for the now great Westinghouse Company at a salary of thirty dollars a month. He was paid this for cleaning machines, polishing brass, and oiling bearings. A good mind for figures soon landed him in the testing department, and one day, in an idle moment, he made upon a sheet of paper in front of him that characteristic mark of the technically educated man—the integral sign of the calculus. That was the moment which Lamme's superior officer chose to stride into his vision. He saw the guilty mark—and the result was not a call-down but a promotion to the design department, which had use for such marks, at seventy dollars a month.

By 1892 the work of harnessing Niagara was soon to begin. In the engineering circles of the day a heavy battle was being fought over the desirability of direct versus alternating current generation—for the cogent arguments of Elihu Thomson had not yet generally won the engineering world. In this fight Lamme said nothing. But he carried a spear, and it was a sharp spear. He showed by test in the Westinghouse shops that generators of equal size could stand 40 to 50 per cent more capacity as alternators than as direct-current machines. The result was a victory so pronounced for Lamme that in this great pageant of power he ceased to be a super and assumed a speaking part. Alternating equipment was, upon his demonstration, installed, following the two-phase alternating-current plan of Nikola Tesla. When, in 1897, the rotors of the first large-scale hydroelectric power development in the world began to turn, they marked a brand-new stage in the development of power.

Having conquered Niagara, Lamme now went on to conquer a more difficult stream—the stream of traffic that had risen in New York City. Until the third year of the present century the elevated trains of New York had been hauled by curious, rectangular and tuberculous cousins of the now large and healthy steam railroad locomotives. In 1899 Lamme and the Westinghouse Company were summoned to develop generating stations such that the Manhattan Elevated trains could thereafter be driven by electric motors and control equipment as designed by Frank J. Sprague. The result was the largest—and one of the last—installation of its kind. Lamme designed eight alternating-current generators driven by reciprocating steam engines. When the engines made seventy-five revolutions per minute the generators would turn out five thousand kilowatts. The generators for this establishment set up a still unbroken record for physical size. They were some forty feet high and the combined weight of stator and rotor was about one million pounds. Had steam engineering, during those years, remained unambitious, the limit of power would have been reached soon after 1899 by reason of the ponderous size which electrical machinery was forced to assume when its rotating parts ran at low speed.

EDISON'S LOCOMOTIVE

Its youthful strength failed to convince the president of the Pennsylvania Railroad

MARKETING THE THUNDERBOLT

A battery of transformers on a high tension line.

MODERN POWER HOUSE

Interior of the 74th Street station of the Interborough Rapid Transit Company, New York. The turbines in the foreground are each from five to thirteen times as powerful as any one of the larger Corliss engines at the rear, yet they weigh less

Fortunately, however, the turbine was now forging to the front in men's minds. Fortunately also, there was, in Benjamin Garver Lamme, an electrical designer not only willing but anxious to utilize those high speeds. Even before the elaborate but prematurely obsolescent plan of the Manhattan Railways in 1899, Lamme had completed another design unique in the history of power—the first in which a steam *turbine* drove an electrical generator. This turbo-generator installation was begun in 1898, when Lamme designed an electrical machine having a six-pole rotating field, to run at twelve hundred revolutions per minute.

It was set in motion by the Hartford Electric Light Company in Connecticut and provided the first faint indication that the reciprocating steam engine of Thomas Newcomen and James Watt had reached its zenith. Decline was to follow swiftly enough. In 1901–02 the alternator was well established and was driven by the steam engine. By 1903 the alternator was beginning to be driven by the steam turbine. Its size was much smaller, its operation easier, its upkeep less, its regulation better, and its speed more constant. For the first time in one hundred and fifty years James Watt had encountered a setback.

Meanwhile the genius of Benjamin Lamme still grew. In 1900 he had become assistant chief engineer of the Westinghouse Electric & Manufacturing Company; by 1903 he was chief engineer. And in 1905 he began upon one of the proudest jobs of his career. The New York, New Haven & Hartford Railroad proposed to electrify its right of way from New York to Stamford, Connecticut, with hints that this electrification might some day extend as far as New Haven. The conventional electrical engineers of the day proposed a 600-volt system; Lamme, caring nothing for convention and everything for efficiency, astounded everyone by proposing an alternating system of 11,000 volts. And

11,000 volts it was. Lamme had triumphed again. When, in 1924, he died in Pittsburgh after a life of sixty years, he had many times over reimbursed the world for the economic loss that it had suffered when the spool he tried to drive too fast as a small boy exploded in protest and pulverized itself over the landscape of western Ohio.

CHAPTER XXI

The New Colossus

As Matthew Boulton himself once said, his partner, Mr. Watt, tended to undervalue his own abilities. He likewise tended to undervalue the abilities of his steam engine. He denied its application to the locomotive. He regarded high pressures as so immoral that he once sought a bill in Parliament to make the use of them a criminal breach of safety. And he indignantly rejected one more improvement which occurred to Jonathan Hornblower in the year 1781.

Hornblower came of a family of the Newcomen engine era. In his home at Penryn, in Cornwall, he studied engine models and suddenly fell heir to an idea ahead of his time—the idea that if steam upon leaving the cylinder of one of Mr. Watt's engines had sufficient energy to expand into the condenser, it might also have sufficient energy to expand into another, larger cylinder, and there do additional work before it expired in the condenser. He applied for a patent. In 1781 he was granted it with the specification of using two cylinders—employing steam in the second after it had done its work in the first and thereafter condensing it by passing it over a metal surface which was in contact with cold water on the opposite side. Thus there came into being the compound engine and the separate surface condenser. Newcomen had condensed his steam by the same means as Hornblower, but had done it in the cylinder; Watt had made the condenser separate but had used a jet, not a surface, for cooling. There were other ingenious specifications to Hornblower's patent but, like Watt's own, it contained no suggestion of form or dimensions which his engine was to take.

Thus protected, however, Jonathan Hornblower began building engines, and the firm of Boulton & Watt soon began to feel his competition. Watt furiously objected; it seemed to him that his own invention of the steam valve cut-off supplied all of the advantages of compounding with none of its defects. He retaliated with a patent suit, and having won it, crushed Hornblower for good and all.

He did not succeed in crushing Hornblower's idea. The Watt patents expired in the first years of the nineteenth century, and in 1804 Arthur Woolf brought the compound engine to life again. Watt had shown that steam at four pounds per square inch expanded to four times its original volume before reaching atmospheric pressure again. Woolf tried steam at five pounds per square inch and discovered that it expanded to five times its original volume. In high excitement he decided that he had hit upon a great natural law.

He was not far from one. To prove his point he began some extensive experiments on the interrelationship of pressure, volume, and temperature of steam. He pushed its pressure up to forty pounds per square inch and to his great happiness discovered that the released steam expanded to forty times its original volume. From this he concluded that the expansive force of steam went far further than had been suspected before—perhaps, he thought, as high as two hundred to three hundred times its own volume.

By 1805 he had built several engines and had a patent on them. Their *modus operandi,* expressed in his own words, was this:

> Let both pistons end their stroke at one time, and let us now suppose them both at the top of their respective cylinders, ready to descend; then the steam of forty pounds the square inch

entering above the smaller piston, will carry it downwards, while the steam below it, instead of being allowed to escape into the atmosphere, or applied to any other purpose, will pass into the large cylinder above its piston, which will take its downward stroke at the same time that the piston of the smaller cylinder is doing the same thing. . . .

By 1815 the Woolf engine seemed a success and in Cornwall the Messrs. I. and J. Lean, mine superintendents, made a number of comparisons between the economy of it and of the engines of Mr. Watt. Woolf had somewhat the better of it—but he had more than the compound principle working on his side. His engines were new and had been built with better tools. Watt's were old and had been built with the necessary crudenesses of their day. For a time the compound engine flourished. Its power was smoother and, what was more, it had a further advantage unsuspected alike by Watt and by Woolf—each cylinder held more nearly to a constant temperature than was possible with the single expansion engine. This was Watt's first principle—yet in the guise given it by another he failed completely to recognize it.

Unfortunately, the compound-engine idea collided suddenly with the high-pressure single-expansion engine of Richard Trevithick, which was much simpler to operate and cheaper to build. This, the famous Cornish Pumping Engine, had a way with it, and soon the compound engine, having been prematurely born, languished in an unfriendly world.

Jacob Perkins in 1827 sought to help matters by patenting an engine of two cylinders—of equal length but one eight times the diameter of the other. He proposed to admit steam to the smallest cylinder at fourteen hundred pounds per square inch pressure and cut it off at one-eighth stroke. To this there were only two disadvantages—there was no boiler that could produce steam at fourteen hundred pounds pressure and no cylinder that would not explode when subjected to it. In England the steam engine began upon a dull twenty years—but in the United States a new genius was rising.

Upon the engineering scene there now appeared one George Henry Corliss, born in Easton, New York, on June 2, 1817. His father was a country doctor and Corliss, having little money or inclination for formal learning, progressed not much further than a grammar school education. He spent an ingenious boyhood in Greenwich, and then, having designed a shoe-sewing machine, he set off to Providence, Rhode Island, to consult the firm of Fairbanks & Bancroft on the details of building it. Machinery had been somehow born in his blood. The quiet of the country was to him much less of an idyll than the clash and noise of a machine-driven city.

Thereafter, Corliss remained a city dweller. Fairbanks and Bancroft were engine builders; they realized the ability of the young man who had come to them with his sewing machine, and persuaded him that his gifts lay not so much in the manufacture of shoes as of power. Corliss agreed with them, and in less than a year, when he was not yet thirty years old, he became a partner of the firm with the idea of a new valve gear for the steam engine already buzzing in his mind.

The engine valves used in this day still resembled the D-valve which had been invented by Watt's assistant, William Murdoch. To Corliss the idea of a rotating valve instead of a sliding one seemed much more sensible—a sort of revolving door to guide live steam in and dead steam out of the cylinders. It would be an improvement, no doubt, but it would also be complicated—with four valves to each cylinder, one in each corner and the connecting wrist plate oscillated by an eccentric from the main shaft. To the valve-releasing gear Corliss attached dash-pots to close the valves quickly and quietly.

Having gone that far, Corliss began a rigorous redesign of the steam engine in every part from frame to flywheel. By February, 1848, when he was thirty-one, he had developed a 260-horsepower engine of 22-inch cylinder diameter and 72-inch stroke. So radically different was it from anything that had gone before that James Watt himself would have had almost no idea that it was an engine. Engineers of the year did not care for it. They smiled amusedly and chanted:

Levers, links, and motions various
Endless jimcracks all precarious.

But used in Corliss's own shop the engine showed great economy. By 1849 Corliss had built and set into operation two more. Now he was busy building a new factory which would take care of his output.

In 1849 Corliss had taken out a patent upon his improvement. Upon that conventional signal for trouble, an engineer by the name of Frederick Sickles, who had taken out a patent upon steam engine improvement seven years before, raised his head. He investigated, insisted that the Corliss improvements had been stolen from his ideas, and brought legal proceedings. These tied not only Corliss, but anyone using his engines, into a number of deft and complicated legal knots. Corliss had to defend not only himself but his customers. In the fifteen years of litigation which followed he was forced to spend over $100,000 before the courts, in their leisurely way, sustained him upon all points.

By 1857 Corliss had revolutionized engineering practice with his cross-compound engine, which finally brought to the principle of Hornblower and Woolf the success it deserved. By 1859 he had taken out six patents upon steam engine improvement—among them a new speed governor to supersede the "ball governor" of James Watt. Watt's governor would operate *because of,* and thus *after,* a change of speed.

It flung its arms gayly about, "hunting" for fluctuations to alter, whereas Corliss's gear actually held the engine at an absolutely uniform speed.

Out of his new governor Corliss made a further advantage. Cotton mills were now becoming users of steam as well as water power. But a cotton mill needed an engine of remarkably steady revolution since the spindles of the mill must revolve at a constant speed and a small variation in engine speed was multiplied many times at the spindles because of their higher number of revolutions. Corliss was not slow to advertise this point.

With these improvements Corliss claimed his engine would now save 50 per cent at the coal pile for an equal amount of power delivered. Having made this boast, Corliss was now forced, like James Watt before him, to back it up with his own funds. People were prejudiced. His valve gear looked complicated, and no one would believe the stories of economy upon which he insisted. So it was that Corliss, like Watt, offered to deliver engines free upon condition that purchasers would compensate him out of fuel saved. For one of his early engines, 180 horsepower strong, installed in a flour mill, the owners agreed to pay Corliss the value of fuel saved over a period of five years. The cost of the engine was about $7,000 and the happy Corliss, his contract signed, eventually received in five installments, $19,734.22. He needed it. The embarrassments of Mr. Sickles were costing him money.

In 1852 he took a contract for an engine to drive the machinery of a rail mill. He guaranteed that it would deliver one third more power on two tons of coal per day than the previous engine had delivered on five. Not only that; he agreed to pay $1 for each pound of coal used by the engine in excess of his guarantee. It was a dangerous bet, but his engine upheld him. Soon he went further; for $7,100 he furnished a 200-horsepower engine to replace one using nine tons of coal per day. Corliss guaranteed that his engine would use but four—and that he would pay $5,000 for every ton it used in excess at the rate of four tons per day. Once again he succeeded.

To a certain extent Corliss was betting upon a sure thing. He was a shrewd business man as well as a good engineer, and he was well aware that the boilers of the engines he was replacing were seriously overloaded. Overworked boilers were uneconomical. Corliss knew that his new engine would not overload them and that therefore better boiler economy was possible. Out of this came an augmented economy in fuel. With the new Corliss engine it was necessary that the boiler generate one half to one third less steam per square foot of heating surface than it had generated before. Furthermore, Corliss was winning now through his realization that superheated steam was a far more efficient source of power than the mere volatilized water of Newcomen and Watt.

It was in 1876 that Corliss built the "Centennial Engine" —the most famous reciprocating engine ever constructed and viewed as an "unsurpassed masterpiece" at the time of the Philadelphia Centennial Exposition. It was a vertical two-cylinder engine of walking-beam type and developed as much as 2,500 horsepower. In Philadelphia, where it moved in the slower tempo of its surroundings, it delivered only 1,400 horsepower at 36 revolutions per minute, but with that it drove every wheel that turned in Machinery Hall. It weighed 607 tons, and on its journey to Philadelphia 35 freight cars were necessary to transport its dismembered sections. Each of the two 27-foot walking beams weighed 11 tons; the 30-foot flywheel weighed 56 tons; the 25-foot connecting rods were forged (the old world was being melted down, now, for the new) from 9,600 shoes ravished from the dead or dying heels of 2,400 horses. For years nothing could approach this engine in power or ambition.

At the close of the Exposition it was sold to the Pullman Company, where it engaged in the pleasant business of supplying power to manufacture sleeping cars until a turbine, pigmy in size in comparison, replaced it in 1910.

For sheer grandeur, the steam engine probably reached its peak with this product of the Corliss genius, even though further refinements were still to come in the reciprocating machine. The compound principle, thanks to Corliss, was now firmly established. But in 1874 it occurred to Dr. A. C. Kirk that if two cylinders were better than one, three might be better than two and—whisper it gently—four might be better than three. The history of the triple and quadruple expansion engines began then. The great John Ericsson inveighed against them in language elsewhere stated, and not until 1881 did they begin to come into general use—usually at sea, where larger and larger transatlantic ships called for concentrations of power not, in the simple 'eighties, required upon land. The boilers of those engines would operate at some two hundred pounds per square inch, and their coal consumption began to touch such economical figures as a mere pound and one half per horsepower per hour. They were beautiful machines of gleaming metal, moving in stately arcs, with all the silent power of a panther poised for a spring.

And they were, too, the ultimate in the development of James Watt's and Thomas Newcomen's idea, for soon the world was to revert to an older idea and out of it evolve the newest in the world of power.

CHAPTER XXII

A RETURN TO HERO

OF ALL the people who remained dissatisfied with the steam engine as improved by James Watt, the most dissatisfied was James Watt. One basic defect in it troubled his engineering mind and made him think that in some ways Hero of Alexandria had had a better idea 150 years before the birth of Christ. Hero, it will be remembered, had spouted steam from radial arms on a copper ball and by that had made the ball revolve. Then, after seventeen and one half centuries Branca, in 1628, had driven a wheel by blowing steam against its vanes.

These ideas had the virtue of simplicity which James Watt's engine lacked. Its greatest defect was that it was intermittent, even after Watt had applied steam to both sides of its piston. Between these pulses of power the piston and its heavy connecting machinery had to slow down, stop, and reverse direction. The great virtues of momentum were violated. But what was to be done about it? No one knew very clearly. Thus, in 1769, in the first year of his triumphs with the Newcomen engine, Watt sought unhappily to replace it by a completely new idea.

For obvious reasons it remained an idea only. Watt thought that by some means, if he used sufficient valves and gadgets, he might force steam to push a weight around inside a hollow wheel; that he could, in other words, chase a piston continuously around inside a doughnut.

> . . . A steam wheel [said Watt], moved by force of steam acting in a circular channel, against a valve on one side, and

against a column of mercury, or other fluid metal, on the other side, was executed at Soho, upon a scale of six feet, and tried repeatedly, but was given up as several objections were found against it.

If any machine were ever predestined to die at birth of excessive friction, this was it. Yet Watt persisted. In 1782 he produced a design for a second rotary steam engine only slightly less absurd than the first. It consisted of a cylinder through which there ran an axle. Upon the axle there was a radial arm which scraped along the inside wall of the cylinder. Into the cylinder steam was admitted at one point with the idea that it would drive the arm and axle around to another point at which the steam was exhausted. A flap valve would keep the intake and exhaust portions of the cylinder separate, and would be thrust rudely out of the way each time the rotating arm went by it. This machine, too, proved a failure, which may be just as well. Had it succeeded, it might have left the world well powered, but stone deaf from the noise of it.

One more keen mind now applied itself to the problem. It belonged to the Reverend Edmund Cartwright, who in 1784 had allowed his thoughts to stray sufficiently from the parish of Goadby Marwood in Leicestershire, to design machinery for weaving—a duty which heretofore had been confined, as the Conservative Party thought it should be, to housewives. Thereafter, this highly secular pursuit absorbed him. He patented a power loom, a wool-combing machine, and began to operate a spinning and weaving mill. He is justly remembered as the father of the textile industry. But by 1793, the worldly equipment of a canon serving him inadequately in the conduct of eighteenth-century business, he surrendered everything cheerfully to his creditors. His mind, however, refused to leave these subjects, and by 1797 he had pushed his thoughts from looms to the engines which drove them. He not only patented a type of condenser quite different from that of James Watt; he likewise designed a rotary engine. It was similar to Watt's second design but an improvement upon it, for it had two steam inlets and three pistons on its rotor instead of only one. The basic unsoundness of the design was not, however, ameliorated by these additions, and the gifted Edmund Cartwright, by now prebendary of the Lincoln Cathedral, resignedly abandoned it, his godliness insufficient to cope with its tremendous friction.

In the last year of the eighteenth century, William Murdoch, Watt's assistant and the same man who, had he been a little less plastic to religious instruction, might have been known as the inventor of the locomotive, took hold of the idea. He helped to construct a machine having two horizontal gear wheels in mesh, side by side. He planned to blow steam through them at the point where they meshed—the gears to run in a steam-tight box with an exhaust opposite the inlet. By the standards of the day his design was perfectly practical; indeed it survives now in some types of rotary pump. But no mechanic of the eighteenth century could build the machine with the necessary accuracy, and once again the idea lapsed.

Then, between the years 1801 and 1830, the world was literally deluged with notions, some from sensible engineers, others from crackbrains. In 1837 Gillman had a really brilliant idea. He devised a rotary steam engine upon a principle which, like Murdoch's, is still in use. Instead of corseting his machine with steam-tight pistons he decided to use steam for its velocity. No one had previously realized how fast steam could travel; its force was thought of only in terms of static pressure. Gillman constructed a series of wheels in a casting walled off by diaphragms between each wheel. Steam, entering one wheel at the centerline, would discharge itself at the periphery, then find its way through the diaphragm into the centerline of the next wheel. The effect of this invention was to add up the force of a number of Hero's Æolipiles. It was designed upon sound principles, and after a fashion it worked. Yet Gillman was hamstrung by the two difficulties which had defeated everyone to date: there were no precision tools in the world and therefore workmanship was not exact. And neither Gillman nor any other practical man had any worthwhile knowledge of thermodynamics—the French physicist Carnot had done his great work but it remained sadly unknown to most engineers.

The idea which all these inventors were worrying and striving over was so simple that one may wonder why Hero of Alexandria himself did not immediately make his Whirling Æolipile into a sinewy machine capable of doing the work of all the engine world. The answer is that although rotating steam machinery works upon a basic principle of utter simplicity, the practical application of this principle quickly begins to involve the most intense complexity. In 1837, when Gillman was groping with his steam wheel, the science of thermodynamics was a philosophical secret locked in the minds of mathematicians like Carnot. To-day the design of a steam turbine needs acres of mathematical symbols and involves the use of such frightening words (no attempt will be made to explain them here) as "adiabatic expansion" and "entropy." This latter word expresses a highly necessary conception for engineers but one so abstract that philosophers in search of the absolute have recently come to take it up in delight at the amount of confusion which it can be made to conceal. If anyone had used either of these terms to James Watt, he would have laughed a harsh Scot's laugh and gone back to business. They are none the less vital to today's steam engineering design.

The engineering world of 1838, however, knew little of them. It was in that year that Heath chanced upon an intensely important idea. Until now it had quite naturally been thought that a steam nozzle and a water nozzle should be the same shape. When water rushes at high velocity from the nozzle of a fire hose it does so in an effort to move from a place of greater pressure to one of less pressure. A converging nozzle increases the velocity of water for the reason that water is practically incompressible, and consequently must travel faster through the narrow nozzle. Steam, on the other hand, being intensely expansive, works differently. What Heath discovered was that a *diverging* nozzle produces a greater velocity for the very reason that it permits the steam to expand. When steam expands it gives up energy which in this case takes the form of increased velocity. Heath understood his discovery imperfectly, but at least he made it. When he did, he produced a right-about-face in steam engineering and made it possible for those who followed him to convert the expansive force of steam into kinetic energy—something that everyone from Hero onward had wanted to do and had done most imperfectly. Everyone had insisted upon choking the steam at the very instant when he wanted it to do its greatest work.

Now the path was clear, yet for another fifty years no one saw the opening in the maze. Then, in 1887, the great sciences of steam engineering and dairy farming collided. The result was the first practical steam turbine to be created after one hundred years of work and one thousand years of hope.

Carl Gustav Patrick de Laval was the agency of this collision. He had been born in Blasenborg, Sweden, in 1845, the son of an army captain. At eighteen he had entered the University of Upsala and later the technical university at Stockholm, from which he graduated in 1866. By 1872 he wore the proud decoration of doctor of philosophy. Then he started his technical career as a mining engineer. He was sent to Hartz, in Germany, to study sulphur production. He gave up mining in 1875 and became a mechanical engineer in the employ of the Kloster Iron Works, with a strongly inventive streak showing beneath the layers of mining engineering which had been laid on him in college. Soon he had developed a new system of air distribution for Bessemer converters and a crucible for the galvanizing process.

Then the idea occurred to him (curious for anyone who had never suffered the drudgeries of farming) that centrifugal force should be put to work skimming cream from milk. He mentioned it to his employer, who played his act conventionally by scoffing. It is hard to scoff successfully at a Swede; De Laval answered by packing up and returning to Stockholm in 1877 to pursue his idea further.

It was not an intensely complex machine that he devised, but it was highly effective and original and separated milk from cream with all the quiet skill of an archangel picking the just from the unjust. In small sizes it could be driven effectively by hand power, but De Laval was thinking in terms of steam for large installations and was bemoaning to himself the fact that the steam engine, after all these years, was still a relatively slow-speed instrument, under the necessity of converting reciprocating into rotary motion. De Laval had a look at the ideas of Hero and of Branca and was heretic enough to decide that they were better. Hundreds of men had made that decision before him—each one of them had come to grief.

By 1882 he had devised a machine based upon Branca's original idea of blowing steam against the blades of a wheel and making them revolve by the impulse. He was a good engineer; his machine had been soundly built, yet its efficiency was discouragingly low. For six years he worked and sweated and studied over his simple plan—almost in despair that such

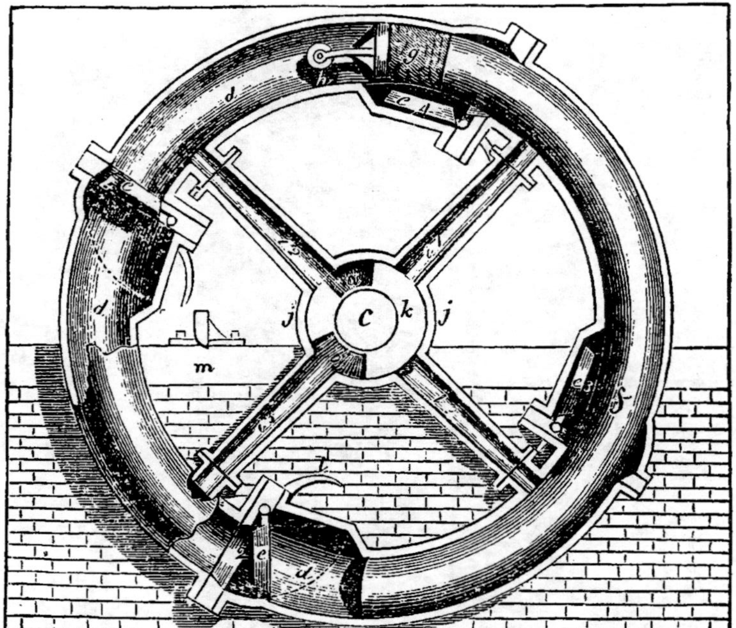

WATT'S FIRST ROTARY ENGINE

IN 1769 Watt hoped that this piston-in-a-doughnut design would supersede his reciprocating steam engine

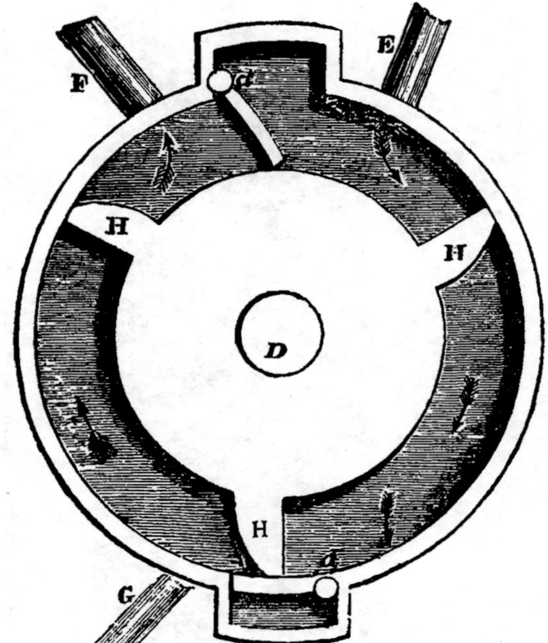

ROTATING STEAM ENGINE

DESIGNED in 1797 by the Reverend Edmund Cartwright, father of the textile industry

New York Central Railroad

"NUMBER 999"

Perhaps the most famous locomotive ever built. In 1893 it hauled the Empire State Express on its famous world-record run of 112½ miles per hour

Baltimore & Ohio Railroad Archives

FIRST ELECTRIC

Placed in service by the Baltimore & Ohio Railroad in 1894, this electric locomotive was the first ever to see service on a standard steam railroad

a beautiful idea should turn out to be such a lamentable machine. But by 1888 the truth had burst on him that all he lacked for an efficient rotary steam engine was the type of diverging nozzle that Heath had chanced upon in 1838. Then, instead of rimming his wheel with rudimentary flat blades to catch the steam, he calculated their curve and surface mathematically and put the mouth of his diverging nozzle so close against them that the steam could have no option but to expend itself enthusiastically in kicking the turbine wheel about.

As a result De Laval got the efficiency he needed, together with all the speed he wanted, and more. Of its nature such a machine was bound to be one of high velocity. If the flywheel of an early steam engine had turned one hundred revolutions per minute even that, after all, was speed which a world used to water wheels that turned twice per minute could comprehend. But now, all of a sudden, an engineer hunting for a way to separate cream from milk had hit upon a steam-driven machine capable of turning itself on its axle from ten thousand to thirty thousand revolutions per minute. It was likewise capable, as De Laval was later to find out, of delivering several hundred horsepower in an extraordinarily compact space—a rotor four inches in diameter (about the size of a horse's hoof) being able to deliver a full horsepower.

Slowly, De Laval convinced unbelievers, including the Patent Office of the United States—which in 1889 refused a patent on the ground that the machine would not work, and only granted him protection for his invention a year after one of his turbines had astounded the World's Fair at Chicago in 1893. He was hot on the track of his idea now, realizing that efficiencies grew better and better as steam pressures grew higher and higher. By 1897 De Laval's boilers were not only supplying power to the Stockholm Exhibition but were running the shops of the now well established

De Laval Steam Turbine Company at Jarla. It was there, on March 9, 1897, that the first commercial use of really high-pressure steam actually began.

The engineering world looked up in vast surprise at what De Laval had done. Each tiny breath of steam against his rotor blade was by itself almost too small to notice, yet De Laval had so arranged matters that ten million or more impulses might be given to his wheel per minute. Such was the velocity of the steam he used that a lead bullet, traveling as fast, would penetrate two feet of oak. But success, as it always does, brought new problems with it. At the tremendous speed which steam could now give rotating machines the necessity for absolute balance was essential; the centrifugal force on a blade weighing only a few ounces could very well reach several hundred pounds. And while the Conservative Party muttered that here was a new force in the world to subvert the ancient virtues, the Liberal Party rubbed its hands in glee at the thought of a brand-new means of releasing energy—a way so good (although one would never have thought to look for it in a milk pan) that had it been successful first the reciprocating engine might still be struggling to be born.

Then the fourth son of the third earl of Rosse entered upon the engineering scene—curiously unhampered in his ideas by having grown up in a moated castle. He was Charles A. Parsons, born in London in 1854. At St. John's College, Cambridge, he had taken a degree as Eleventh Wrangler in 1876. He was a sportsman, too; rowed in the first boat of his bumping club and won the college pairs. With such a background it would have been only natural had he devoted himself thereafter to the most strenuous cultivation of a life of ease. But his father before him had been a man of science; thus, perhaps, it was considered neither surprising nor dishonorable that his young son apprenticed himself, in 1877, to the firm of Armstrong & Whitworth, the English gun and battleship designers.

Here Parsons studied under the genius Whitworth and turned his attention to the propulsion of ships. In a bare six years he emerged, at the tender age of twenty-nine, with a turbine new and elaborate in principle which soon proved to be one of the greatest single inventions in all the history of power. By 1889 Parsons had his own works at Heaton; from which were to emerge those amazing machines tremendous in power and watch-like in delicacy which were to revolutionize marine propulsion and win for their inventor most of the medals and honors that scientific men can wear.

Whereas De Laval's turbine acknowledged Branca as its father, the turbine of Charles Parsons was a direct return to Hero, for in it not impulse but reaction was the principle involved. That is to say, the steam, being permitted to expand only in the moving blades of the rotor, kicked the rotor around much as the vanishing bullet from the muzzle of a rifle kicks the stock against the marksman's shoulder. Moreover, Parsons knew that one set of blades could never extract sufficient energy from high-pressure steam. His manifest answer was not one wheel but many—and the multi-staged turbine of the present day was thus born and patented in the year 1884. It consisted not of one wheel but of a large number of discs strung in a row along a central shaft and inclosed in a steam-tight casing. To these wheels there were fastened blades to catch the steam; between them, attached to the casing, there were stationary blades to guide the steam on a longitudinal path through the casing. The steam thus writhed through the turbine, first striking moving, then fixed, blades until it had gone through many stages and in its efforts to escape had driven the rotor eighteen thousand revolutions per minute.

But the accelerating wonders of this idea, at once the oldest and newest in the history of power, were not to end with Charles Parsons—very shortly and rightfully Sir Charles Parsons, for his great and honorable contribution to the mechanical world. The next skip which the turbine took, having already jumped from a milk pan to a moated castle and the University of Cambridge, was from the University of Cambridge to Brooklyn, New York. Here Charles E. Curtis, an electrical manufacturer whose industry had helped to make life worth while through the agency of the electric fan, had interested himself in turbine development. Out of the riches that had come to him he began after 1895 to investigate the idea of combining the virtues of the De Laval and Parsons turbines into one machine.

It was a noble idea but it was expensive. Curtis did not have cream separators in mind when he built his machines; he was thinking of ocean liners. It was thus not at all difficult for him and his associate Crocker to spend sixty thousand dollars in experiments at his Brooklyn factory. For some time no one paid much attention to them, and into the turbine they continued to pour most of the profits from their more practical ventures.

Finally, however, the General Electric Company, which had by now come to realize that the life mate for the generator was the turbine, bought the Curtis idea and began to experiment upon that grand scale which, in a newly complicated world, only great corporations might afford. The General Electric Company spent over three million dollars on the Curtis turbine before it became a commercial success. But when it did, the association of the reciprocating engine with the generation of electricity, so hopefully begun by Thomas Alva Edison in 1882, began to draw toward a close. On land Curtis turbines now produce fifteen million horsepower; at sea they produce twenty million and do it in a space which would almost put James Watt's original "Old

Bess" to shame. Its development, once the experimental stages had gone by, was prodigious. In 1903 a Curtis turbine drove a 6,700-horsepower generator in Chicago and the world gawked at the marvel. To-day turbine units of 100,000 horsepower excite no attention at all. These turbine units are the equivalent of some 3,600,000 slaves—yet the combination of boiler, turbine, generator, and condenser needs only some 10,000 cubic feet instead of the 173,000,000 cubic feet which an equivalent number of slaves would require.

But even to-day the turbine retains, along with its vast power, that watch-like delicacy which is the constant concern of engineers. A turbine rotor seldom revolves less than two thousand times per minute—and as it does, its blades are only a tissue paper's width removed from the intricate stationary parts which direct its energy. With dimensions such as these, turbine engineers do not always rest easy. A few drops of water are enough to raise havoc with their machinery. For a turbine is very similar to an elephant, which any mouse can make squeal. Every so often, despite the best of care, something mysterious happens inside the rotor casing and, to the accompaniment of the horrid sound of ripping steel, some fifty thousand dollars' worth of delicate machinery tears itself to pieces. Yet these accidents are surprisingly rare and the turbine's amazing advantages of efficiency, lack of vibration, and absence of any necessity for internal lubrication make it an indispensable machine in to-day's demands for power.

Parsons, De Laval, and Curtis had opened the eyes of the engineering world to some new facts. A turbine four feet long was quickly developed into a prodigious power producer: something the size of a baby elephant now had the strength of three thousand horses. So efficient a heat machine did it become that it would extract ten degrees of temperature from the steam with every inch that it traveled through the casing. And this new concentration was but a reflection of the startling properties of superheated steam. Steam at 150 pounds pressure per square inch, passing through a nozzle to atmospheric pressure, travels 4,000 feet per second, or over 45 miles per minute. It puts to shame the muzzle velocities of shells from the largest guns: a fifteen-inch shell has an initial velocity of but 2,900 feet per second—equivalent to a sedate pace of about 30 miles per minute.

It is this which may indicate why water is the great enemy of a steam turbine. Water has 125 times the mass of steam at 200 pounds per square inch; therefore, drops of water moving with the same velocity as the steam which entrains them are amply capable of shattering steel. A machine whose parts are boiler-turbine-condenser is a curious combination. To develop 70,000 horsepower the boilers must evaporate 300 tons of water per hour, and the condensers must change an equal amount back to water again. Yet this water, until it puts on the cloak of invisibility which fire will bring it, is an enemy, not a friend, to the machinery of which it is the vital blood.

CHAPTER XXIII

The Sweating Thrust-blocks

It was obvious enough to Sir Charles Parsons that his turbines were destined to go to sea at an early age. Their compactness, economy, and lesser vibration argued mightily in their favor. The Conservative Party, on the other hand, viewed them with great alarm. By 1890 it was quite clear that the turbine was soon to be another annoying innovation in marine engineering—equally discommoding with that produced by John Ericsson, when he banished the paddle wheel in favor of the propeller.

During Ericsson's day, and thanks largely to his influence, steam was making great strides at sea. Yet no ship, up to the middle of the nineteenth century, had ever been built so large as that craft three stories high and made of gopher wood which had been launched upon the shores of the Red Sea when the rains descended and the waters prevailed. Noah's Ark was 300 cubits long, 50 cubits wide and 30 cubits high; since a cubit is about 18 inches, Noah's Ark must have been a formidable craft 450 to 500 feet long.

Lacking the direct injunction from God, no modern marine engineer had been sufficiently courageous to build a competing ship, until, in 1845, Scott Russell and his associates began, upon the banks of the river Thames, the construction of that premature leviathan known as the *Great Eastern*. She was 692 feet long on deck, 83 feet in the beam (118 feet over the paddle boxes), 58 feet in depth, and 28 feet in draft. Her displacement was some 28,000 tons. It took four years to build her. To build, that is, but not to launch. Lest she run into the opposite bank, it had been decided to put her into the Thames broadside-on instead of stern first, and although the cost of launching is usually 10 per cent of the cost of the vessel, the original company backing the *Great Eastern* went bankrupt in trying to get her into the water. All in all it took three months, after the first attempt, to persuade her into the element for which she had been designed.

She was a famous and picturesque craft, and a triumph of naval architecture. She had a double bottom (its first appearance), and 50 water-tight compartments. Her owners boasted of her as the safest ship ever built to that time—a boast which she proudly justified when once, despite a frightful gash which she gave herself on Fire Island, she made port under her own steam.

Her engines had been built by the now ancient and honorable firm of James Watt & Company, Birmingham. There were eight of them—four to drive the two 56-foot paddle wheels, four more to drive the centerline propeller. Into the bargain, the *Great Eastern* was a full-rigged ship, and in one particular marine engineering reached its zenith with her: no other ship in history has ever had five funnels and six masts.

Her paddle engines had cylinders 74 inches in diameter, working at the enormous stroke of 14 feet. Her propeller engines were almost as vast: their cylinders were 84 inches in diameter and they worked at a stroke of 4 feet to drive a screw 24 feet in diameter. Her boilers had 44,000 square feet of heating surface—something over an acre. Yet, for all these tremendous physical dimensions, she was tragically under-engined for her size and could turn up little more than 8,000 horsepower to drive a hull of 28,000 tons displacement. She lived an unhappy and frustrated life, and in 1887 there seemed nothing to do with her save turn her over to the ship-breakers and scatter her twenty anchors to the seven seas.

Despite this setback, steam navigation went on to further triumphs. In 1855 it had taken 2,200 tons of coal to carry 800 tons of freight. Fifteen years later it took 800 tons of coal to carry 2,200 tons of freight, and Sir Lyon Playfair could make the proud estimate that:

> A small cake of coal which would pass through a ring the size of a shilling, when burned in the compound engine of a modern steamboat, would drive a ton of food and its proportion of the ship two miles on its way from a foreign port.

The English now had risen from the horrible agony of the Industrial Revolution and were well on the way toward the height of their proud, imperial steam-driven destiny.

So things went, by slow development and painstaking refinement. Hulls were being made of steel, not iron. Auxiliary sails were being furled, on merchantmen and men-of-war, for the last time. Twin-screw ships were becoming commonplace. The old single-expansion engines were giving place to compounds. Everything was quiet, well organized, and pleasant. Then, at one stroke, the development of the turbine suddenly threatened to turn marine engineering upside down once more.

In 1894 a syndicate was formed in England to exploit the great development of Sir Charles Parsons. The inevitable resistance of the Conservative Party took the form of pointing out that the turbine was an irreversible machine and that to drive a ship astern another complete turbine installation would be necessary. Furthermore, the speed of a turbine was, at lowest, from 2,000 to 4,000 revolutions per minute, whereas the maximum efficient propeller speed was only 300 to 500. To reconcile these two speeds there would have to be heavy steel reduction gears.

Nevertheless, timidly argued the Progressive Party, a turbine offers much more power in a single unit than does a reciprocating engine which, for all its refinement, is a bulky, unwieldy, and highly vibratory machine. Space is important upon shipboard; give us a little time and we will overcome the turbine's disadvantages.

Without much encouragement the syndicate, between 1894 and 1897, constructed and equipped the ship *Turbinia.* To drive her a radial flow turbine of 1,500-horsepower was installed and connected to a single-screw propeller. On trial the *Turbinia* developed the speed of 18 knots. But the pioneer difficulties of working with new machinery broke out in every direction. The *Turbinia* was far from a happy ship, and the original turbine was ignominiously removed.

To replace it (the syndicate had the courage of desperation now) not one but three turbines were installed. They were piped in series so that the steam at its highest pressure, after leaving the first turbine, would enter an intermediate one and from that pass to the low. To accommodate these changes the *Turbinia* was now equipped with one centerline propeller shaft and two wing shafts on each side. One turbine drove each shaft and upon each there were three propellers. A reversing turbine was coupled to the low-pressure stage on the centerline shaft to answer calls for backing. With this installation, which totaled 2,400 horsepower, the syndicate was rewarded by watching the *Turbinia* travel faster than any ship had ever traveled before; she shot through the water at 34½ knots, whereupon the Conservative Party capitulated even unto its highest command—the British admiralty.

As a result of this surrender the British navy began building two destroyers, *Viper* and *Cobra,* to be fitted with turbines. They were diminutive ships of 370 tons displacement.

In their enthusiasm for producing speed the contractors kept the hulls feather light. Both ships were launched in 1899 and both immediately came to melancholy ends. The *Cobra*, after a successful trial trip, went to sea and, lacking enough longitudinal strength, promptly broke in two. The *Viper*, no more fortunate, crashed upon the rocks at Casquets and likewise broke in two. Turbines had in no way caused either accident, yet as a result there came now a pause of several years in any further adaption of the turbine for work at sea. This was doubly unfortunate, since the *Viper* before her catastrophe had proved herself the fastest ship in the world. She could travel 36.86 knots, or barely four less than the destroyer record of the present day.

The navies of the world are always alive to the necessity of speed. The British admiralty proceeded with turbine installations in first- to third-class cruisers and then in that famous ship, the *Dreadnought*—whose name is now the generic term for all modern battleships. The *Dreadnought* was launched from the Royal Dockyards in Portsmouth in 1906 with Parsons turbines driving four screws, each shaft being coupled to an ahead and an astern turbine. Her speed on trial was 21 knots—some two to three knots faster than any other battleship. The superiority of the turbine was now so decisively indicated that the British admiralty adopted it for all capital ships—a bare nine years after the *Turbinia*.

The same year saw the United States navy make its first timid experiments with Curtis turbines upon the cruiser *Salem*. Results were pleasing, but the navy yearned for one more test. When the sister battleships *North Dakota* and *Delaware* were launched in 1908 and 1909, the former was equipped with Curtis turbines and the latter with reciprocating engines. The results so favored the *North Dakota* that thereafter the reciprocating engine at sea became, after a bare ninety years of life, an obsolescent machine for large ships.

Even before this, merchantmen had cast envious eyes at the new equipment. The Allen Line was the first to screw up its courage. In 1904 it equipped its ships, the *Victorian* and the *Virginian*, each over 10,000 gross tons, with turbines which would supply 12,000 horsepower. That inspired the Cunard line which, in the following year, built the turbine-driven ship *Carmania*. So satisfactory was she that Cunard thereupon determined not only to equip two new ships with turbines, but to make these ships the largest and fastest in all the world. Whereas 30,000 horsepower had hitherto been considered the limit, each of these new craft would have four turbines of 17,500 horsepower each—a total of 70,000. It was a mammoth conception. There would be some 3,000,000 turbine blades for each ship—driving shafts which would weigh 250 tons. Twenty-three double-ended boilers and two single-ended boilers would be needed, and the sweating stokers would have no less than 192 furnaces to feed. From those plans, determined in 1906, sprang the *Lusitania* and the *Mauretania*.

They were the ultimate word. For two years these sister ships fought with each other for the north Atlantic speed record. The *Mauretania* finally captured it in 1909 with a crossing, between Ambrose Light Ship and Cherbourg Breakwater, of five days, ten hours, and forty-one minutes. Soon thereafter came the war and the sinking of the *Lusitania*. When the Peace Treaty was signed at Versailles, marine engineering had apparently spent itself and was resting upon its oars. Ships grew more luxurious, but they grew no faster. Twenty years after her launching, the *Mauretania* still held the blue ribbon of the sea.

Then, in 1929, ensigns dipped and whistles thundered to mark a new event. The *Bremen* had broken the *Mauretania's* record with a crossing of four days, seventeen hours, and forty-two minutes. This was in July; in October she was to break her own record by a matter of eighteen minutes more, and in March, 1930, her sister, the *Europa,* was to reduce it to four days, seventeen hours, and six minutes.

In the engine rooms of the *Bremen* and the *Europa* the art of the turbine designer has reached new peaks of complexity. A trip to the engine rooms is bewildering; surrounding the vast mounds of cream-colored metal, which are the four main turbines, there are such wealths of auxiliary machinery, control apparatus, electrical generators, condensers, pumps, piping, and cables that it is almost impossible to see the forest for the trees. Most of the moving machinery is so thoroughly concealed that the engine rooms seem mere heaps of static metal. Only the enveloping heat and the constant rushing noise make you realize that here, in a space no larger than a triplex apartment in New York, there is compressed the power of more than one hundred thousand horses. No longer separate machines as on the *Turbinia,* the three stages—high, intermediate, and low—of a modern turbine are all geared to the same shaft, with a fourteen-ton bronze propeller spinning at the far end of it. As always, backing turbines must be provided, but these, unlike the ahead turbines, consist of high and low stages only and have but 60 per cent of the power of the ahead turbines.

All this complication describes only one turbine—and there are four. With four turbines of three rotors each for going ahead, and four turbines of two rotors each for going astern, there are, to drive the ship's four propellers, a total of twenty distinct pieces of main propulsion machinery. The ahead turbine rotors must revolve some sixteen million times during the course of one Atlantic crossing—always with a tissue-paper clearance between moving and stationary parts.

Query one of the engineers in charge of these monsters and

he will tell you that turbines, for all their vast power, still remain so delicate that his crew must always, at least metaphorically, keep them wrapped in cotton for their complete comfort. There must be, for example, a small auxiliary reciprocating engine attached to each shaft to keep the rotors turning over slowly all the while that a ship is at rest with steam up. This is because the rotor of a hot turbine, not revolved, would expand excessively where the heat of the intake pipe warmed it, and in an hour would be so completely lopsided that any attempt to start it would be disastrous.

Is this, then, the end to complexity of power at sea? Not at all. Even before the World War, engineers were experimenting with a new idea which now, for naval vessels, at least, is perhaps destined to supersede all others. That is the so-called turbo-electric drive designed by W. L. R. Emmet of the General Electric Company.

In the utilization of this idea the United States navy came first. It made the first large-scale experiment in 1913, upon the U. S. Collier *Jupiter*. Electric motors on this ship of 19,450 tons displacement drove it at fifteen knots with a fuel consumption 25 per cent less than that of similar vessels using reciprocating engines. In principle, the electric drive is simple enough. The turbines, instead of being connected through gearing to the propeller shafts, are directly connected to dynamos, and these dynamos, through cables and control apparatus, drive motors which, in their turn, propel the tail shafts.

But why go to the trouble of converting mechanical into electrical energy only for the sake of converting it back to mechanical energy again after a few feet of transmission? On navy ships, at least, there are a number of excellent advantages. For going astern the driving motors may be reversed by throwing a switch, while the irreversible turbines continue to drive the generators in the same direction. In the second place, space is saved through the elimination of long sections of propeller shafting with their bearings and couplings, as well as by the elimination of the heavy reduction gears. The turbines may be placed near the boilers and the motors near the propellers. Better economy is possible because turbines should run constantly at design speed. There is an even greater flexibility of control, since, when running at cruising speed, all four propulsion motors can be turned by one turbo-generator set. Last, no reversing turbines are necessary, so that not only is space saved but full power is always available for going astern. Naval engineers claim this further advantage, for battle purposes, that the separation of vital parts of the engine room in far distant sections of the ship makes it much more unlikely that one or two high explosive bursts could completely destroy the machinery. Since, however, no modern dreadnaught of the United States navy has ever been engaged in battle, that belief still remains unbaptized by fire.

The first electrically driven battleship of the United States navy was the *New Mexico,* equipped in 1917, whose 31,200 horsepower drives her at 21.1 knots, and which can make 17.5 knots with only one generator in action. The generator's 2,037 revolutions per minute can be reduced to 168 propeller revolutions per minute without gearing. Minus auxiliaries, the propelling machinery weighs but 476 tons.

Even merchantmen are turning, cautiously, to the principle. The three ships of the Panama-Pacific Line, the *Virginia,* the *California,* and the *Pennsylvania,* which travel from the east to the west coast of the United States via the Panama Canal, are electric driven, with two turbo-generator sets on each ship driving two propellers with one motor each. Each vessel develops 17,000 horsepower.

To-day the most tremendous concentration of seagoing power in one unit is found in the United States aircraft carriers *Lexington* and *Saratoga.* Each of these ships develops 180,000 horsepower and can travel faster than 30 knots. And so highly designed is the apparatus that from the switches in the central station one man can control all speeds of the ship. To each propeller shaft (there are four in all) there are two direct-connected induction motors each running at 5,000 volts and delivering 22,500 horsepower. Their stators are 15 feet in diameter. At full power the motors drive the propeller shafts at 317 revolutions per minute. Power is supplied by four turbo-generator sets. So vast is its concentration that the power for the blower motors used to cool the main motors and generators, together with the power supplied to run all pumps, is more than the total power generated on board the super-dreadnaught U. S. S. *Colorado.* The pumps aboard these ships have a fresh and salt water capacity of over twenty-one million gallons per hour—enough to supply the constant needs of a city of four million people.

Here electricity has reached its most complex and concentrated power development. On each ship there are more than 600 auxiliary motors—ranging from small half-horsepower units for ventilating equipment, up to the 400-horsepower motors to operate the anchor windlass and steering gear. So tremendous is the power of the *Saratoga's* generators that when, some years ago, a water shortage threatened to cut off the electric power of the city of Seattle (population 315,652) the city made formal request upon the Navy Department that the *Saratoga,* then anchored in the harbor, might be coupled to the mains of the city. Her 180,000 horses went from ship to shore and were more than ample to supply the entire city's needs.

It is to such heights that power at sea has risen. It is a great change that the seas have seen in a little over a century. In 1812 the frigate *Constitution* was the proudest ship of the United States navy, and upon her sails, so later calculation

Science Museum, London

THE *GREAT EASTERN'S* ENGINES

A WORKING MODEL of the power plant which drove its paddle wheels

Science Museum, London

THE *GREAT EASTERN'S* ENGINES

THESE drove the propeller which supplemented the paddle wheels

Science Museum, London

EARLY STEAM TURBINE

THE historic design of Sir Charles Parsons, through which men's ideas of power and speed were reborn

showed, there pressed a force equal under average conditions to 600 horsepower. It was in that same year that Michael Faraday doomed the sails of the *Constitution* and made possible the machinery of the *Saratoga* by his first experiments with the voltaic pile, made even before the end of his apprenticeship to Mr. Riebau, bookseller of Blanford Street, London.

CHAPTER XXIV

SPENT WATERS

It is very seldom that any painter sufficiently gifted to have his works hung in the Louvre is also a specialist on centrifugal pumps, universal joints, rope and belt drives (open and crossed), spiral and bevel gearing, and link chains. But those were only a few of the subjects with which one painter concerned himself while at the same time he was busy with the painting of La Giaconda and The Last Supper. The painter was Leonardo da Vinci, with whom in many ways the modern world began.

Among the adjuncts to life which Da Vinci sought to improve was the water turbine. In his incredible career he found time to design a horizontal impulse water wheel—not very good, it is true, but markedly better than any horizontal wheel had been before him. It was a simple wheel with curved vanes convergent toward the center. When water struck them they reacted by a quiet but continuous revolution.

Somewhat better was the pit wheel first shown in the sketches of Besson in 1568. It had a cone-shaped rotating wheel placed in a masonry pit so that water in striking its vanes acted both by impact and by gravity. The water, having done its work, escaped at the bottom of the masonry cone —but if the rotor and its pit did not fit closely much of the water and hence much of the power was lost. The sixteenth century having reached that impasse, nothing more happened until a new scientific curiosity made its appearance in about the year 1725.

It was known as Dr. Barker's Mill, yet it was in no sense a mill, and was not even Dr. Barker's; its principle was a direct reversion to the æolipile of Hero of Alexandria. Dr. Barker wrote his name in history by the simple device of using water instead of steam and causing two radial pipes to revolve around a central axis that was hollow and full of water. The pipes revolved by the reaction set up when the water spurted from holes in the end of the radial arms. Most lawn sprinklers to-day are just so many Dr. Barker's Mills.

The same Dr. Robison who was a friend and counselor of James Watt studied this hydraulic curiosity and said,

> The unbalanced pressure that makes the wheel revolve is equal to the weight of a column of water having the orifice for a base and twice the depth under the surface of the water in the trunk, for its height

which now seems slightly more complicated than an explanation of a lawn sprinkler need be. The science of hydraulics was new, and learned natural philosophers like Désaguliers, Euler, Bernouilli, and De la Cour worked earnestly to bring a useful machine out of the idea. If, De la Cour thought, a large pipe were brought down from an elevated reservoir, its lower end bent upward and topped with one of Dr. Barker's Mills, upside down, the result might prove to be a water motor. This apparently was as far as he could progress in 1775.

For years this simple machine teased the imaginations of inventors. In 1845 the Scotsman, Whitelaw, of Paisley, suggested what he thought to be a spectacular improvement but which came down to nothing much more than bending the two radial arms so that they formed a letter S. This calligraphic improvement was quite unnecessary. Someone else had meanwhile invented a real water turbine, and hydraulic engineering took a sharp upturn.

The inventor was Benoît Fourneyron, who began experiments in 1823. In 1827, at the age of twenty-five, he completed his first turbine. The Romans had achieved one great step in the conservation of energy when they narrowed their rivers so that the full stream struck the paddle wheels of their barges; M. Fourneyron accomplished another when he enclosed his wheel in a complete, water-tight casing so that every drop of water which he fed it could, like Ralph Waldo Emerson, find no escape from duty but performance. Water entered this motor at the center and diverged in every direction to the periphery of the wheel, with a success which greatly exceeded Fourneyron's own expectations.

Fourneyron's next turbine was not built until 1834. It was for the decoration of the iron works of M. Carron at Francho-Comté, where its duty was to compress air to blow a blast furnace. It could produce 7 to 8 horsepower and was willing, obedient, and tractable. It could find nourishment even in a head of water as small as 9 inches, hitherto useless for ordinary water wheels. The enchanted M. Carron thereupon gave Fourneyron a mammoth order—he was to deliver a water turbine of 50 horsepower to replace two old water wheels which together could equal only 30. This 50 horsepower must be wrung out of a stream with a head of but 4 feet 3 inches. Fourneyron delivered this difficult machine, saw it work with complete satisfaction, and thereafter was deluged with orders. For his successes he was kissed on both cheeks by the Society for the Encouragement of the Arts and had pressed into his palm the sum of six thousand francs as a prize for the construction of the then best large-scale horizontal water wheel. Although his first machines were less efficient than the large lumbering vertical wheels, they were much more compact and, like their early horizontal predecessors, they needed no troublesome bevel gearing to connect them to millstones.

By 1840 the use of the Fourneyron turbines had become widespread in France. The Conservative Party, no doubt, clung to the old mill wheels with the assurance that what was good enough for their great-great-grandfathers was good enough for them. But visions of great power were in the air now, and France was well supplied with these new and spectacular machines.

In about this same year Yankee ingenuity first asserted itself in the matter of hydraulics when one Uriah A. Boyden, a New England engineer, applied himself to water turbines and succeeded in going M. Fourneyron one better. His improvements were confined to details but they were important, and when, in 1846, the mills of the Appleton Company at Lowell, Mass., were equipped with Boyden wheels, it was discovered that these mills were getting the benefit of an 88 per cent efficiency—perhaps the best that had ever been then achieved by any type of hydraulic equipment.

To the mid-nineteenth century, when electrical experiment was being pursued by men of science and laughed at by gentlemen who called themselves "practical," the new efficiency of the hydraulic turbine was a gift from the industrial gods. But one of its first results was a tendency to remake cities like Lowell in the image of Venice, with canals crisscrossing the town to supply water to the various factories. To the literal New Englander, however, a river was a river and a street was a street, and he had no wish for the mental confusion that resulted from finding one where he expected the other. Thus arose transmission problems, which appeared in several guises. The first was the mechanical transmission of power by shafting—but this was limited to distances of three hundred or four hundred feet. Pulleys and wire rope were an improvement. One could stretch a rope 540 feet between pulleys before it sagged too greatly; and could transmit power a little over a mile by using ten spans of pulleys, before friction began to destroy results. And where the contours of the terrain did not interfere, the energy of the waters might, at the central point, be bottled up into pipes and thus under pressure be delivered to the rotors of the turbines.

So schemed the world of 1870. But the once infant science of electricity, thanks to the genius of Michael Faraday, had been weaned from its barbaric diet of ha'pennies and sulphuric acid.

By now, both hydraulic power machines and electrical generators having reached an adult age, their parents, the water wheel and the friction machine, soon married them off. A small hydroelectric plant sprang up at Meisbach in Bavaria and transmitted current through an iron wire to the city of Munich. In the United States, a few weeks before the date in 1882 when Edison's Pearl Street station was opened, a hydroelectric plant had begun operation at Appleton, Wisconsin, and had lit 280 lamps.

It remained for an Ohio carpenter who had gone to California during the gold rush to demonstrate the great uses of hydroelectric power. In 1884 Pelton decided that gold lay in water, not as nuggets but as power. He built a water wheel to be operated by the waste surface water of the Chollarshaft in the Sierra Nevadas. He had a unique natural development with which to work. The head of water was 1,600 feet and it was, therefore, Pelton's necessity to develop some type of water wheel which would utilize this tremendous fall. Pelton met the issue by directing a thin stream from a small nozzle at tremendous velocity against a bucket shaped a little like two halves of an eggshell held side by side. The ridge in the center of the bucket split the stream which, breaking into two parts, swirled around inside the bucket and reversed its direction, transmitting about 80 per cent of its energy to the wheel, on the periphery of which a number of these buckets were placed. To this wheel Pelton hitched a dynamo of 130 horsepower designed by Charles T. Brush. With his installation he was able to drive (not all at once, of course) six 80-horsepower electric motors at the Nevada Stamp Mill, over a mile away. This, in the simple days before practical alternating current or transformers, was an achievement which made him famous. To-day the site of Pelton's experiments is included in a great power system over eight hundred miles long in which 114 hydro-electric stations and 30 steam plants are interconnected over 7,200 miles of high-tension wire and deliver a total of more than two and one-half million kilowatts.

The Pelton wheel remains to-day an indispensable machine where high heads are to be used, and thus finds one of its greatest theaters of development in Switzerland. At Lake Tanay, an effective fall of water 3,018 feet high serves a Pelton wheel so well that it can extract some nine horse-power from every quart of water it uses. The velocity of the water at the nozzles is approximately two hundred miles per hour. The generators attached to this wheel develop a 5,000-volt current for distribution in the Bouveret district. Even this development was later exceeded by that on Lake Fully, where the highest head in the world—5,412 feet—supplies a nozzle pressure of 2,260 pounds per square inch.

Perhaps it was Pelton's demonstration which turned the envious eyes of engineers once again toward Niagara Falls, where thunderous waters, falling 159 feet, provided a magnificent source of potential power. The first attempt to harness these waters, which had been going to waste since the world began, was made in 1861, when a canal was built to divert a flow from the rapids above the falls. The Civil War prevented any utilization of this design until 1877, when some small wheels were installed to supply power a short distance by shafting.

By 1891 electrical generation and transmission was a practical art, and in that year the famous power house Number One was begun at Niagara to house ten 5,000-horsepower turbines which would operate two-phase alternating current generators. When this power house began to function in 1897 it marked a new stage in the development of power. Since that time there has been an almost continuous enlargement of power houses on both sides of the Niagara River, and the waters led into the turbines through canals of solid rock generate current which is transmitted more than one hundred miles.

Even to the Father of Waters, the modern engineer pays no respects. The shoulders of the Mississippi River are now in harness at Keokuk, Iowa, where a dam almost a mile long produces a 32-foot head of water and gives up 300,000 horsepower from 30 turbines. Current at 110,000 volts is delivered as far as St. Louis, 144 miles away. Fully as important is the Wilson Dam at Muscle Shoals on the Tennessee River in Alabama, which will some day develop a capacity of 112,000 horsepower.

These new giants of power dwindle into insignificance by contrast with the Hoover Dam now in construction in Boulder Canyon of the Colorado River. Here a dam 730 feet in height (nearly twice that of any dam yet built) and 650 feet thick at the base will hold back a lake of water 115 miles long and 582 feet deep. In this will be stored 30,500,000 acre-feet of water—enough to cover the entire state of New York to the depth of one foot. The power plant will generate a minimum of 660,000 horsepower—and more during the period of high water in the river. The power plant will be housed in two buildings, one on each side of the river and each containing six hydroelectric units. Here will be generated more power than in the entire state of California, or in the state of Illinois, including Chicago.

Yet it is still not enough. In the United States 11,700,000 horsepower is developed from falling water, but four times as much as that sidles away to sea unused. Engineers have calculated that the world power available from water is 240,000,000 horsepower, of which but 30,000,000 is now developed. It is obvious enough that the story of hydraulic power has not yet properly begun and that some day unborn generations may look back at the Hoover Dam with the same faint smile of condescension which we, in this present day, bestow upon Dr. Barker's Mill.

CHAPTER XXV

Prairies and Mikados

By the time the Civil War in America was over the railroads were great. An artery spanned the continent. Promoters were busy extending capillaries throughout the countryside in the happy belief, not to be dashed for almost half a century, that a railroad would make money no matter where it ran. Locomotives had reached practicable limits of speed, trains were steadily becoming heavier, the volume of passenger and freight traffic was rising almost day by day. Swooning in the embrace of a great era of post-war prosperity, railroad promoters cared very little that the operation of their roads was far from safe, and that the number of accidents grew and grew to proportions greater than the timid House of Lords in 1820 could have dreamed of.

For one thing the locomotive of the 1860's was a bull in a china shop. It was easy to start, and easy to maintain at high speeds; but urged on by the momentum of a string of heavy cars behind it, it was difficult to stop. Hand brakes supplied the only means available, and it was not unusual when fifteen minutes elapsed between the time that the engineer whistled to his crew for brakes and the time when those brakes brought the train to a halt.

It was not until the very year in which the transcontinental railroad was completed that one George Westinghouse, a young man of twenty-three, completed his first design for a curious locomotive auxiliary. By means of it he proposed, to the amused scorn of the Conservative Party, to stop a train with wind. Westinghouse called his device the air brake, and his first design was of equipment now known by the odd title of "straight air." In it, air compressed by a steam-driven pump on the engine was conveyed under the cars of the train by a pipe and acted directly upon pistons which pressed brakes against the wheels. This mechanism, wonderful as it was for its time, had some serious defects. The engineer found it impossible to apply the brakes simultaneously on all cars, since the air, under obligation to push pistons as it went along, took rather a long time to travel backwards through the train. Unbraked rear cars tended to pile up against braked forward cars. Even more serious, if a hose connection broke at any point the brakes would not operate.

These defects in his system young Westinghouse began to overcome in the year 1872 with the design of that fascinating and complicated instrument known as a "triple valve." The braking system was now entirely different. The engine's air pump charged separate air tanks under every car, and the pressure in these tanks would apply the brakes only when the triple valve permitted it. The triple valve was actuated by the engineer's valve in the cab; when the engineer moved his lever for an application of the brakes the pressure in the brake pipe was *reduced,* and this reduction in pressure opened the triple so that air was applied to the brakes from the individual tanks under each car. The greatest advantage of the new system was that a break in the hose line would now, instead of rendering brakes inoperative, automatically set every one and bring the train to a stop. But not until 1887 did Westinghouse, by his invention of the quick-action triple valve, succeed in making his system comfortable as well as safe. Until that year wise passengers, when the brakes went on, tightened their belts, took off their spectacles, and clutched what was nearest them.

With its fatal defect of uncontrollability eliminated, the locomotive was ready for a new spurt forward in design. The firm of Matthias Baldwin was busy with the construction of locomotives and in 1867 had built the first of a new type designed to haul heavy freight upon inconvenient grades. It had three pairs of drivers and a two-wheel pony truck in front. It was named the "E. A. Douglas" and was of a type which later became famous as the Consolidation. This type hauled most of the heavy freight in the United States for the next thirty years.

Meanwhile, locomotives began to differentiate further, and engineers busied themselves with abstruse calculations of tractive efforts, adhesion, grate area, steaming capacity, and other details that had never bothered their predecessors. It was now quite obvious that a locomotive to haul a fast express should be as different from a locomotive to haul a heavy freight as a race horse from a Clydesdale. Type began to follow type in great profusion.

The confusion was equally great. Engineers could settle upon no satisfactory means of classification until, at the Master Mechanics' Convention at Saratoga, N. Y., in June, 1901, F. M. Whyte, mechanical engineer of the New York Central & Hudson River Railroad, proposed that all locomotives be classified by their wheel arrangements. It was a wonderful idea, and those now great companies, the Baldwin Locomotive Works and the American Locomotive Company, before long adopted it officially. It requires but three numbers, usually separated by dashes—the first to indicate the number of wheels under the pilot, the second the number of drivers, and the third the number of trailers under the cab. Although it was not thought of until 1901, let us apply it now to all locomotives of whatever age and thus derive the advantages of its irresistible simplicity.

The locomotives built in the United States prior to 1850

were usually 2–2–2 or 4–2–0. For some now mysterious reason early engineers insisted upon using no more than two driving wheels, thus seriously limiting the tractive efforts of their machines. But after 1856 a type was developed, known as the Mastodon, the wheel arrangement of which was 4–8–0. It was not, however, until 1861 that the first famous type came into being—the Mogul, or 2–6–0. Five years later the Consolidation, or 2–8–0, had been developed. A few simple principles began now to be increasingly observed. A freight locomotive needed a large number of drivers to give it enough adhesion to pull its heavy loads. One of the earlier examples of this design was the Decapod, or 2–10–0, developed for heavy freight in 1885.

For slow speed only two pony wheels were needed under the pilot, but at higher speeds it took four to guide the locomotive safely over any combination of curves and switches without danger of derailment. Furthermore, lighter loads required only four drivers; but when economics demanded the use of cheaper fuel, a larger fire box became essential. The fire box could not overhang too far without additional support. An extra pair of drivers, although they would add to adhesion, would have the disadvantage of making the fire box high and therefore narrow and difficult to fire. Thus it was that small trailer wheels under the cab became a necessity. They appeared on the Columbia, or 2–4–2, in 1892, and on that famous passenger type, the Atlantic, or 4–4–2, in 1895.

By 1897 designers found it necessary to produce still more tractive effort and still better steaming capacity. The latter requirement demanded an increase in boiler and grate area. The result was a 2–8–2 design called the Mikado, because Japan, of all countries, requested a locomotive for narrow-gauge freight service more powerful than the Consolidation and likewise capable of burning an inferior coal. Inferior coal meant a bigger fire box: the Baldwin Locomotive Works, taking the model of the Consolidation, increased the fire-box size so much that it then became essential to add trailing wheels behind the drivers to support it. The happy Japanese, unaware of the American impiety which had named this new steam monster for their sacred ruler, used Mikados in large numbers to haul freight, while American railroads sat calmly by for five or six years, continuing their use of the Consolidation.

But in 1902–03 American roads were driven to demand a freight engine which would burn lignite. Baldwin answered by repeating the 2–8–2 design, this time in standard-gauge sizes. The greater power and speed of the Mikado soon began to show up the defects of the Consolidation. The Mikado proved its abilities by hauling fifty cars loaded with wheat from Buffalo to Philadelphia, 416 miles, in twenty-seven hours.

By 1900 the Mogul, or 2–6–0, was being supplanted by the 2–6–2 type known as the Prairie. It was little more than a Mogul to which a larger fire box had been added, thus producing the necessity for trailer wheels. It was designed by Baldwin and took its name from its first use—which was for hauling freight on the monotonous, level western lines of the Chicago, Burlington & Quincy Railroad. It was popular with Western roads for a long time, but passed out of use largely because of the continuing differentiation of models. The Prairie might be used either for passenger or freight service but lacked the crisp, clear-cut advantages of engines designed exclusively for one service or the other. From the Prairie, however, there sprang in 1901 the 4–6–2 type known as the Pacific, still enormously popular for heavy but fast passenger runs.

The Pacific locomotive was first built by Baldwin for the New Zealand Railway. (It is curious how often a new

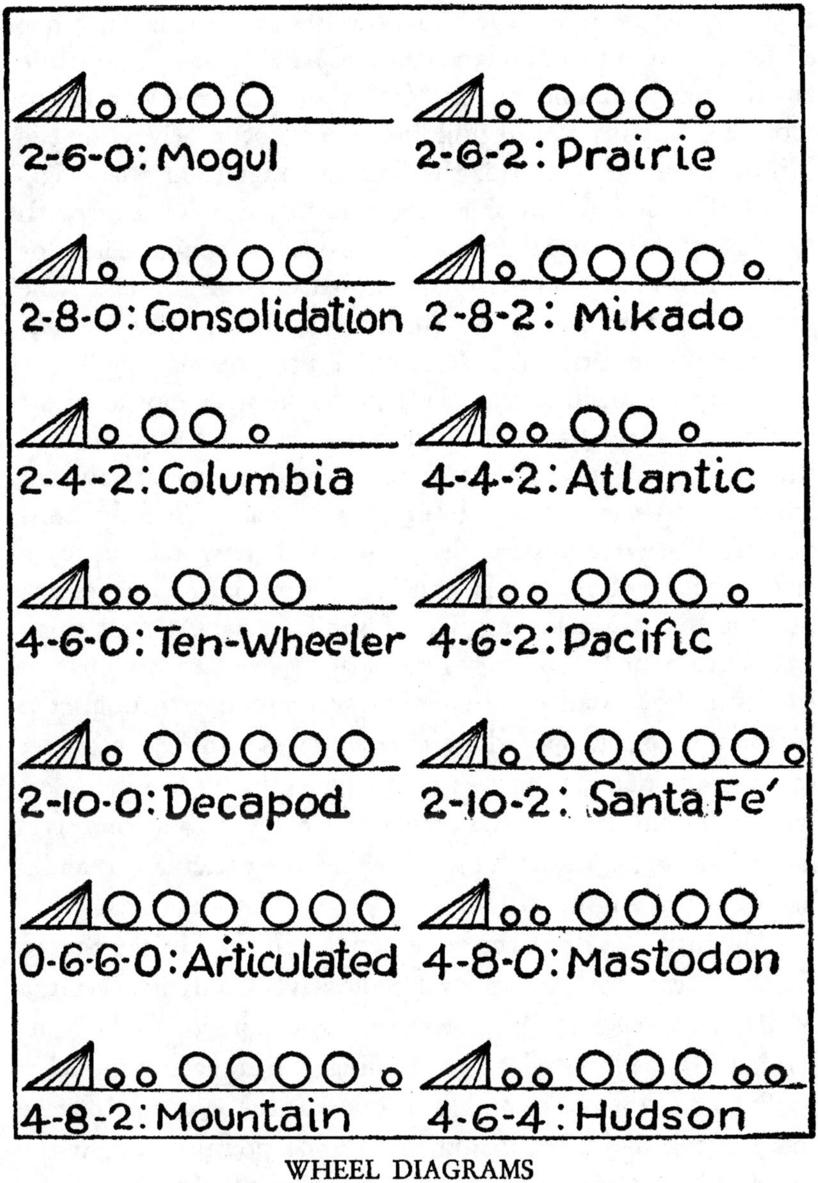

WHEEL DIAGRAMS

THE system devised by F. M. Whyte which classifies locomotives according to their wheel arrangements.

type was developed in the United States at the insistence not of American but of foreign railroads.) Baldwin had no name for this new machine, but in 1903 when the American Locomotive Company began building 4–6–2 locomotives for the Missouri Pacific Railroad, the Pacific designation was originated. The design combined the speed of the Atlantic with the heavy-duty capabilities of the earlier ten-wheeler, or 4–6–0. It was a quick steamer and had a large fire box. The first such Baldwin machine, for a weight of 161,504 pounds, would develop a tractive effort of 18,350 pounds.

But by no means all of the new locomotive development was taking place in the United States. Even as early as 1890 the French engineer, Mallet, had developed an articulated locomotive—that is, one built in two parts, flexibly connected. The virtue of the design was that it would supply a large tractive effort, yet be capable of negotiating curves impossible to a rigid locomotive of similar size. At first small sizes were built for military railroads; by 1900 approximately four hundred Mallet articulateds were in use through the mountains of Europe where sharp curves and heavy grades made some such arrangement essential. Usually the cylinders were compounded, for the Mallet was really two locomotives in one—a high-pressure engine behind the articulation and a low-pressure engine in front.

Not until 1904 did the type first reach the United States. In that year the American Locomotive Company built a Mallet of 0–6–6–0 design for the Baltimore & Ohio Railroad. It was exhibited at the Louisiana Purchase Exposition at St. Louis in 1904 and was one of the marvels of the mechanical world. Its weight of 334,500 pounds made it the heaviest locomotive in the world. Its tractive effort of 71,500 pounds and working pressure of 235 pounds of steam per square inch were equally notable. It could haul a two-

Smithsonian Institution

THE "JOHN B. JERVIS"

A GIANT freight-hauling consolidation of the modern day

IRON PERCHERON

This powerful freight locomotive, built for the Northern Pacific Railroad to burn a low grade of fuel, has thirty-four wheels and weighs over a million pounds

New York Central Railroad

THE HUDSON TYPE

$90,000 worth of motive power, one of a family of 174 locomotives which haul the crack trains of the New York Central Railroad

Canadian Pacific Railway

STEEL SINEWS

IN 1932 this compound locomotive of the Canadian Pacific Railway was one of the largest and most powerful in the world

thousand-ton train at ten miles per hour and demonstrate other feats of strength.

Prior to the introduction of the Mallet into the United States American engineers had sought to solve the problem of heavy freight traffic in their own way. Their answer was the Santa Fe. It was of 2–10–2 design and thus a descendant of the early Decapod, or 2–10–0, with the addition of trailer wheels. Like the Mallet, the Santa Fe was compounded; but its cylinders, instead of operating separate sets of drivers, were arranged in tandem. The two valves on each side were fastened to one stem so that only two sets of link motion were required for the four cylinders.

In the year 1908 South Africa once again inspired American locomotive design by demanding an engine which could cope with the mountains in Natal. Once again a new type was developed from an old by the simple expedient of increasing grate area and compensating for the added weight by adding trailer wheels behind the drivers. This time it was the Mastodon, or 4–8–0, which was transmuted into the Mountain type, or 4–8–2. It developed more tractive effort than its cousin, the 4–6–2 Pacific, and when it was put in service in 1908 in South Africa it was so satisfactory that twenty-five such engines were immediately built. Yet it was not until 1912 that the United States embraced it. In that year the first of the Mountain type was built for the benefit of the Chesapeake & Ohio Railway, which had on its hands a difficult wrestling match with those stubborn barriers to travel of any sort, the Alleghany Mountains. The Mountain type swiftly justified its existence when, on 180 pounds per square inch of boiler pressure, it proved itself capable of making seventy-two miles per hour with ten cars on level stretches. The horsepower it developed was 2,480, and it was well adapted to climbing heavy grades.

Thereafter, experiments with wheel diagrams (save for the

construction of increasingly tremendous freight engines) paused until after the World War. When Europe had partially recovered from that convulsion, M. Asselin, of the Chemin-de-Fer du Nord, in France, felt the necessity for more powerful locomotives to haul the famous Baltic Express. Once again the solution lay in the construction of a larger fire box, and once again therefore trailer wheels were added behind the drivers. This time, however, not two wheels but four were added, and the Pacific type locomotive, having sprung from the Prairie, now sired the Baltic, or 4–6–4. In it Asselin got what he wanted—a locomotive which would haul a four-hundred-ton train at seventy-five miles per hour on level stretches, and do it with a boiler pressure of 227 pounds per square inch, using compounded cylinders. The 4–6–4 design was slow to be adopted in the United States, but in 1927 the New York Central Railroad put into operation the first of the famous "5200" locomotives of that design to haul its crack expresses like the Twentieth Century Limited. With new adaptations the Baltic was now called the Hudson.

This same year, 1927, which saw the introduction of the Hudson locomotive in the United States, was a famous one for locomotives. It likewise saw the construction of a giant Consolidation, or 2–8–0, named the "John B. Jervis," built to haul freight on the Delaware & Hudson Railroad. It is a compound, weighing 335,500 pounds. It operates at 400 pounds per square inch boiler pressure and, unlike almost all other locomotives, has a water-tube, not a fire-tube, boiler. (Water-tube boilers are not usually preferred for locomotives since they are better fitted to deliver a continuous supply of steam, whereas the locomotive, with its starts and stops, has varying necessities.)

Even more famous was the locomotive built in England in the same year—a 4–6–0 type known as the "King George V." At the time the most powerful locomotive in England, it has 78-inch drivers and operates at 250 pounds per square inch pressure. It weighs close to two hundred thousand pounds—but it is not by reason of these statistics that this locomotive is an interesting design. It has four cylinders, two inside and two outside—the inside cylinders being connected by crank axles to the two forward drivers, and the outside cylinders conventionally connected to the two rear drivers. On its trial trip the "King George V" traveled 155 miles, hauling 410 tons at 61.7 miles per hour. Moreover, this is one of the few locomotives which have ever made a round trip across the Atlantic; it was brought to the United States in 1927 for the famous Fair of the Iron Horse which marked the centennial of the Baltimore & Ohio Railroad. As a souvenir of this trip "King George V" now proudly sports a bell upon its bumper—the only locomotive in England (perhaps out of consideration for the hunters of foxes) to carry such an ornament.

To-day it is the Canadian Pacific Railway which possesses the largest locomotive in the world. It is nearly one hundred feet long and weighs nearly four hundred tons. It has, in its trials, pulled a load weighing 7,961 tons and has shown a 30 per cent fuel economy over other freight engines. It is an oil-burning locomotive of the three-cylinder type, which means that in addition to the two outside cylinders of most non-compounding locomotives this one possesses an inside center-line cylinder connected to the driving axles by a crank. It is one of the most spectacular examples of the new tendency toward "multi-pressure engines"—i.e., engines in which steam is generated in three separate boilers and at three different pressures. The fire box is used to produce steam at about 1,350 pounds per square inch. By a complicated system of recirculation this steam is used to generate other steam at 850 pounds pressure, and this in its turn finally generates low-pressure steam at 250 pounds pressure.

It is, however, in Europe, the home of the locomotive, that the most interesting steam-locomotive experiments still continue. Germany is working at the present time with a locomotive operating at pressures from 1,400 to 1,700 pounds per square inch. England has its own particular child in the so-called "Hush Hush" locomotive, a precocious machine developed in 1929–30, upon the 4–6–4 plan. This locomotive is different from almost all others ever built in that instead of casting exhaust steam prodigally up the stack, it delivers it instead to a condenser. This engine looks like no other ever before constructed, since it is completely sheathed in streamline metal from pilot to tender.

Turbine locomotives have been a subject of railroad conversation for many years, but have never yet been found completely practicable. A turbine is pretty much a constant-speed machine, whereas it is the life and duty of a locomotive to start and stop frequently. In Sweden, however, the turbine locomotive has had some success. During the war the brothers Frederick and Birger Ljungström designed a locomotive in Stockholm upon the model of which several larger turbine locomotives, ranging from 1,750 to 3,000 horsepower, have since been built. In this locomotive the boiler is in the leading section, with a trailer carrying the turbine, reduction gear, and condenser. The great disadvantage of this design is that most of the total weight is not placed upon the drivers and that accordingly adhesion is poor. On the other hand fuel economy is excellent. The condenser permits longer runs without additional feed water. There is no such thing as a dead center for a turbine. It supplies a large starting torque, and produces no unbalanced impact to pound the rails.

Yet the greatest rival of the steam locomotive is its cousin, the electric. The first of all such radical machines was originally put into operation in the year 1891 by that consistent pioneer in American railroading, the Baltimore &

Ohio. In 1891 a 650-volt direct-current engine of 4,000 amperes rating was built to haul B. & O. trains through the Baltimore tunnel, thus ridding it of the horrible miasma of smoke and cinders which passengers had hitherto endured. The example was swiftly imitated abroad in railway tunnels under the Alps.

But not until 1907 did electricity begin to come into its own on American railways. In that year the New York Central Railroad completed its plan to provide electrical traction for the thirty-three miles out of New York City, to Harmon, New York. This direct-current installation at one stroke eliminated smoke, noise, and railway yards from the center of the city and at the same time made it possible for the railroad to use efficient locomotives capable of traveling seventy-five miles per hour.

Even more spectacular, however, was the answer of the New York, New Haven & Hartford Railroad, which shared with the New York Central the facilities of the Grand Central Terminal. The New Haven announced several innovations—that, for one thing, as soon as its tracks diverged from those of the New York Central it would operate its trains by overhead wires instead of third rails and would utilize alternating instead of direct current at a voltage, prescribed by Benjamin Lamme, of 11,000. At first the electrification was pushed only as far as Stamford, Connecticut; later it was extended the entire distance to New Haven, with the result that an electrified *mileage* of 72 became the equivalent of a *trackage* of almost 600 miles. By 1928 the New Haven Railroad had achieved a record—its 41 electrical locomotives of the first type, in service for twenty years, had rolled up the tremendous mileage of 51,250,000.

In addition to the New York, New Haven & Hartford, the pioneers in electrification were the Great Northern Railway and the Chicago, Milwaukee & St. Paul. In 1909 the Great Northern electrified its original Cascade Tunnel with a three-phase system, since extended to cover the new Cascade Tunnel and the section between Skykomish and Wenatchee, Washington. It was, however, the Chicago, Milwaukee & St. Paul Railroad which electrified the greatest mileage. Having electrified 646 miles of main line, including the Cascade Mountain and Rocky Mountain divisions, this railroad undertook to test the comparative power of steam and electricity in a primitive but spectacular fashion. It matched, in the arena of power, an electrical locomotive, weighing 520,000 pounds, against a Mallet steam locomotive of 556,000 pounds. Head to head, coupled together to avoid damage, the two locomotives met like antagonistic goats. There was snorting and blowing from the Mallet and utter silence from the electric locomotive. It was the latter which triumphed; the continuous torque of the electrical motors, combined with better weight distribution on the drivers, won the day for it.

It was a victory soon to be reflected in many places. By 1914 the Pennsylvania Railroad had electrified two important divisions running out of Philadelphia. To-day its ambitious plans for electrification call for the expenditure of over one billion dollars and will make the railroad electrically operated over all important trackage lying between New York and Washington.

One great advantage of electricity over steam is gained by reason of the 4 foot 8½ inch gauge from which American railroads suffer. Locomotive boilers long ago reached the clearance limit of width. To the length of boilers there exists a limitation less definite but equally harsh. The limitation is length beyond which heating gases cannot be pulled through the boiler tubes. It is this which, in the last analysis, limits the length and strength and power of steam locomotives and insists that they shall not very greatly exceed the records of the present day. The electrical locomotive on the other hand knows no such limitation. As many units as necessary may be coupled together, and horsepowers may be piled upon horsepowers with no consideration whatever for the problems of clearance and dimensions. And one man may control as many units as desired.

For all the inefficiencies of the steam locomotive it remains a much cheaper machine to operate than the electrical locomotive, which must be fed by elaborate and tremendously expensive systems of stationary steam-plant generators, transformers, transmission lines, and so on. Yet the steam locomotive, gruff, picturesque servant of mankind that it has been, is perhaps, alike with the reciprocating engine at sea, headed for neglect, obsolescence, and death. To-day, in Europe and America, many a steam locomotive rests inactive upon a pedestal of fame in some railway terminal. The time may come some day when steam locomotives of most sorts, from "Puffing Billy" to the proudest Baltic, will be similarly regarded by a generation accustomed to float only upon the liquid and invisible stream of electrical power which is now pouring from the country's generators.

CHAPTER XXVI

The Mad Disciples of Christiaan Huygens

The ability of a certain type of mind to remain resolutely dissatisfied with things as they are has been of remarkable utility in the world of power. James Watt was dissatisfied with his own steam engine because of its wasteful necessity for converting reciprocating into rotating motion; out of his dissatisfaction there was some day to come the turbine.

The now forgotten Robert Street was not only dissatisfied with Mr. Watt's engine but with the whole silly (as it seemed to him) idea of obtaining power from overheated water. Why, he asked himself, should the real source of energy, which was coal, be used so far away from the real working unit of the engine, which was the cylinder? Why not burn the fuel in the cylinder and avoid these inefficiencies of energy transmission? In 1794 an English patent was issued to Robert Street to cover this idea. The engineers of the day smiled long, sad smiles (there are times when erudition is a great disadvantage) and the more intelligent they were the longer and sadder their smiles became. For to them, well grounded in the history of power, Robert Street's idea was a distinct retrogression. The earliest steam engines had used the same vessel for boiler and cylinder; it was only when Newcomen and Watt differentiated these parts that the engine amounted to anything at all. Why now should Robert Street in England and Lebon in France, some five years later, be so anxious to throw away all the progress of a century that lay between Thomas Savery and James Watt?

Yet in making this proposal Robert Street had done nothing new. His idea was slightly over a century old. For the steam engine itself had stemmed from an idea which the Dutch physicist, Christiaan Huygens, had suggested in 1680. In that year it occurred to him that the explosive force of gunpowder might be made to do work if a small charge were placed in the bottom of a closed vessel, the top of which was movable. The gunpowder would explode, lift the piston beyond the point where an exit pipe would open and relieve the pressure—whereupon the piston would fall again. Here the great mind of Christiaan Huygens broke down. The idea of continuous work may never even have occurred to him; if it did, he was obviously bewildered to know what else to do save lift out the piston and insert a new charge of gunpowder. Accordingly he did not build his machine—yet it was this same idea which Dionysius Papin utilized when in 1690 he adapted it to boiling water instead of gunpowder and thus astounded the Royal Society.

On and off, for a great many years, this idea of burning fuel in the cylinder of an engine recurred. Usually such designs were spoken of under the slightly sinister but entirely descriptive name of "explosion engines." Mostly, they were favored by those visionary gentlemen who believed that man, despite the trouble he had propelling himself with any speed on the ground, might eventually fly. Among these was Sir George Cayley, who wrote in 1809:

> The French have lately shown the great power produced by igniting inflammable powders in close vessels, and several years ago an engine was made to work in this country in a similar manner by inflammation of spirit of tar [benzene]. I am not acquainted with the name of the person who invented this engine, but from some minutes with which I was favored by Mr. William

Chapman, of Newcastle, I find that thirty drops of oil of tar raised eight cwt. to the height of 22 inches; hence one horse power would consume from ten to twelve pounds per hour, and the engine itself need not exceed fifty pounds in weight.

Thereupon Sir George gave this extraordinarily straight tip to engine designers:

Probably a much cheaper engine of this sort might be produced by a gas-light apparatus and by firing the inflammable air generated with a due portion of common air under a piston.

If anyone had had the vision and the skill to follow up this idea he could have built a workable internal combustion engine any time during the next week. But in the slow-moving, slow-thinking early nineteenth century, the man who was eventually to follow up Cayley's suggestion was not to be born for another thirteen years.

He was Jean Joseph Etienne Lenoir—born at Mussy-la-Vielle, in Luxembourg, on January 12, 1822. He had gone to Paris in 1838 with no resources and had become a waiter in a café; with the high ambition burning within him of becoming a decorator and enameler. By 1856 he had somehow fallen under the spell of machinery; by 1860 he had produced a "hydro-carbon motor" (the fuel for which was benzene, discovered in 1825 by Faraday) which he fondly hoped might develop three horsepower. It was of the non-compressing type and its lamentable inefficiency together with its excesses of hysteria discouraged its inventor. The obvious necessities of design which Lenoir was overlooking were swiftly suggested by Schmidt in 1861 and Alphonse Beau de Rochas in 1862, who so corrected matters that operations inside the cylinder were made now to follow the regular cycle of intake, compression, explosion, and exhaust which has never, since that day, been very much altered.

Still none too certain of his engineering, Lenoir in 1863

built another engine of a non-compressing type, mounted it on wheels, and (here was the first gasoline-driven carriage in the world) traveled from Paris to Joinville-le-Paris and return, taking an hour and a half each way. It was a success of a sort, but it left Lenoir discouraged. He had accomplished it half a century too late for any public applause; the idea of mechanical locomotion which had shocked the world of 1800 was now used as a commonplace convenience by the world of 1863. Sadly agreeing with his critics that his curious engine could not compete with steam on land, he next tried a six-horsepower engine in a boat. Finding that equally unsatisfactory, he gave up his experiment altogether.

But Lenoir had builded better than he knew. By 1876 the German, N. A. Otto, first built the four-cycle engine that Schmidt and De Rochas had suggested more than a decade ago. And in the United States, George B. Selden began to think about vehicles to be driven by an internal combustion engine. Selden shrewdly saw the scheme of things into which the internal combustion engine would fit. On a grand scale it was not likely to compete with steam. But its compactness in relation to its possible power and the lack of any necessity for carrying about a tank full of water and heating it with an open flame made it, in Selden's eyes, the perfect means for small-scale locomotion. In 1879 he applied for a patent. The Patent Office, working at lightning speed, granted his application in 1895, and the stage was set in the United States not merely for the emergence of that strange freak, the automobile, but for a great and bitter legal tangle.

Meanwhile, in Germany, a student who had spent from 1852 to 1859 in the Polytechnic School at Stuttgart became in 1861, at the age of twenty-seven, an engineer for the Karlsruhe Machine Works. His name was Gottlieb Daimler. By 1872 he was the director of a gas-engine works at Deutz near Cologne, which, during the years of his management, was to grow into the famous establishment better known as the Otto Engine Works. By 1882 he had removed to Cannstatt near Stuttgart, and was giving his entire attention to developing a light-weight internal combustion engine.

In this year the internal combustion engine was of two types: one, utilizing natural or artificial gas; the other, some sort of volatile fluid like benzine or "petroleum spirit"—for the refining of petroleum to produce kerosene had now begun. In either case, however, the gas was ignited by a flame burning in the end of a pipe threaded to the cylinder. In 1883 Daimler made a great contribution to this weak-lunged and uncertain child of power. He invented the system of "hot tube" ignition. Instead of introducing the flame directly into the cylinder, he kept it burning constantly under one end of a porcelain tube, the other end of which was plugged solidly into the cylinder. It was now the red-hot porcelain which ignited the charge of fuel and air; by experience Daimler discovered that he could so closely regulate the temperature of his tube that he could force the explosion to take place at any desired point during the compression stroke of the piston. This was so satisfactory that Daimler decided the ultimate in internal combustion engines had been reached. He could now drive an engine at 800 revolutions per minute instead of the 200 revolutions per minute hitherto possible. For the same weight he could produce almost four times the power. Daimler felt that the time had come to apply this new engine to a moving vehicle.

But another compatriot, Carl Benz, had anticipated him. After graduating from the Baden Technical School, Benz at the age of twenty-eight had opened a shop of his own in 1872. By 1880 he had begun to commercialize a two-cylinder, stationary internal combustion engine. In 1883 he organized his own business, and in 1884 he produced his first moving vehicle. As quietly as possible he experimented with it in private.

Then, in 1885, the quiet streets of Mannheim were suddenly profaned by a strange vehicle. To the ear it gave the impression, not of a means of transport, but of a regiment doing slaughter with small arms and an occasional field piece. Turning horrified in their tracks, sedate residents could see, through clouds of smoke and an occasional spurting flame, a three-wheeled juggernaut approaching. The two rear wheels were vast nets of piano wire connecting thin tires to small polished hubs. The front wheel, for steering, was only slightly smaller; to it was connected a long upright, with a handle for the pilot. High up upon an upholstered carriage seat sat the madman Benz, his feet resting upon wooden slats and his hand clutching the steering handle. Somewhere behind him his gasoline-driven instrument of hell coughed and choked and thundered and, by means of a chain and sprocket, drove the rear axle to which were fastened the enormous wheels.

Despite the barbarous sound of his new invention, Benz had actually produced an engine of tremendous, if not quiet, refinement. In the first place he had superseded Daimler's hot-tube ignition with something greatly superior: into the walls of his cylinders he had led two wires with a tiny gap between them, and by means of a battery, a coil, and an interrupting switch he created the first electrical ignition system ever used. More than that, his cylinders were water-jacketed and the three-quarter-horsepower engine drove a flywheel to smooth out the pulses of energy.

There was no such thing, of course, as a gear shift. The engine was connected to the wheels by a belt transmission which led to a differential and from there to the two chains running to the wheels. Upon all this Benz received a patent in 1886. By 1888 he was amazing visitors to the Munich Exposition, and one year later his firm, for the manufacture of these queer machines, was a stock company with a capital of three million marks.

Daimler, meanwhile, had been far from idle. In the year of Benz's Mannheim triumph he had produced (his soul is doing penance for it now, no doubt) the first motorcycle—then a mere engine-driven bicycle. Daimler heated his porcelain hot tube, vigorously turned the crank of his engine, and on November 10, 1885, the streets of Cannstatt, as those of Mannheim earlier, were treated to a startling barrage. Daimler threw the lever which tightened the belt connecting engine to wheel, and was off upon a journey which was to lead him to renown and riches. The great Daimler Motoren Gesellschaft was born that day. The motorcycle was a popular idea from the start. Within forty-five years the Daimler plant was to see the number of such vehicles rise from one to the impressive total of over 2,750,000.

Europe continued busily now on the improvement of the internal-combustion machine. The great William Maybach, working for Daimler, had designed the so-called aspirating carburetor and supplied something to act as a clutch. In Paris in the late 'eighties the engineer Lavassor, using the Daimler patent, built the first of all automobiles with an engine in front. And in 1892 Panhard and Lavassor first equipped the auto with pneumatic tires.

At last the United States seemed to realize that it had been passed in this strange, smelly, noisy race. At Chicopee, Massachusetts, one Charles E. Duryea, an experienced bicycle designer, worked for the Ames Manufacturing Company. In the summer of 1892 he and his brother, Frank, spurred by these tales of success in Europe, decided to build a gasoline machine which, with a strange flair for morbid etymology, they called a "buggynaut." It was finished in September, 1892, and tested indoors to avoid the ridicule which the brothers were shrewd enough to anticipate. But the engine was too weak; further tests had to be deferred until the summer of 1893.

It was then that the Duryeas completed a second machine, which to their hysterical pleasure made seven or eight miles per hour when it would make anything. A long cast-iron crank was inserted into its vitals from the rear and churned vigorously. When the first detonation announced that the engine was running, it was well not to tamper with it; there was no way to regulate its speed. Theoretically it would not stop again until the gas tank ran dry or the electrical circuit was broken; in actuality the ingenious thing found many more reasons than those. But for all its weaknesses it was a pioneering machine in the United States and stands to-day, as it deserves to, in the Smithsonian Institution in Washington, credited with being the second United States gasoline car. The Duryea brothers gave up bicycle designing and embraced the automobile as a profession. In 1895, only three years after their first failure, they had reached a production schedule of nine cars per year; by 1896 this had leaped forward to ten.

Meanwhile there were competitors. For one there was Elwood Haynes, in 1890 field superintendent for a natural gas company in Kokomo, Indiana. In this capacity he found it necessary to make long trips with a horse and buggy; during these periods of boredom he thought idly about such substitutes as steam, electricity, or gasoline. By 1893 he had purchased a small gasoline motor. He set it up on a block and studied its peculiar anatomy and physiology. Soon he had sketched a vehicle to which he hoped it might be attached.

To Elmer Apperson, a small machine-shop operator in

Kokomo, he took his drawings. From them Apperson, with the help of his brother Edgar—by divine Providence a bicycle repair man—built the vehicle which Haynes had drawn. It had its first trial amid pardonable public excitement on July 4, 1894, and in the patriotic fervor of the day no one jeered. The Haynes machine was pulled to an outlying point by a horse. Thereupon, with Haynes at the wheel, the motor was cranked and triumphantly brought the carriage home again, entirely horseless, at a speed of eight full miles per hour.

There entered now upon the scene a young man whose name was Henry Ford. He had been a severe disappointment to his father. He had refused the gift of a forty-acre farm offered to him on condition that he leave the supposed flesh-pots of Detroit and spend his time in the happy pursuit of agriculture. At about the time when the Duryea brothers were doing their work in Chicopee, Henry Ford had gone to work for Mr. Edison's company in Detroit and was to become, not long thereafter, its chief engineer. Neither steam nor electricity, however, could distract him from the idea of the gasoline engine; he built a two-cylinder one in his spare time with a flywheel made of wood and a cylinder constructed out of a section of a gas pipe. Then he put it aboard a carriage and worked for forty-eight hours without sleep to get the last details of it complete. At two o'clock of a rainy April morning in 1893 he finished it, called his wife, rolled the car into an alley that led to Bagley Avenue, and drove off. It was a short drive. The father of over twenty million Ford engines stopped sullenly in a matter of a few minutes when the nut of a bolt came off. But the idea behind it went on forever—to what amazing lengths it is unnecessary to particularize here. To-day, of all the power generated in the United States, 75 per cent comes from gasoline engines, most of them driving automobiles—with

Smithsonian Institution

HORSELESS CARRIAGE

IN THE late '90's this design by the Duryea brothers helped introduce America to the automobile age

Smithsonian Institution

SELDEN PATENT

MODEL of a self-propelled vehicle by one of America's earliest automotive pioneers

Smithsonian Institution

ELWOOD HAYNES'S BUGGY

ANOTHER early American conception of horseless travel

mammoth central power stations contributing but 5 per cent to the total.

The automotive business in the United States was now established. By 1895 Ford had driven his car over one thousand miles—in continuous trouble with the police almost every mile of the way. They could not discover the violation of any definite law, but their instincts told them that some variety of tort was being committed. Horses felt the same emotions, but the *fin de siècle* small boy did not. So great was the boy's enthusiasm that Ford and other early drivers of horseless carriages had to chain and padlock their cars to lamp posts when they left them standing.

In 1898 Alex Winton set up another milestone by selling the first gasoline-driven automobile to be marketed in the United States. Before the first of the new century companies had been organized in Boston, Philadelphia, and Chicago for the manufacture of automobiles. Their total capitalization had reached $388,000,000—not all in cash. At least eighty were turning out varieties of that vehicle which had, less than two decades before, been the terror of the Mannheim streets. An early transcontinental trip failed, but the automobile of 1899 at least distinguished itself by traveling 720 miles between New York and Cleveland over what, in that day, passed for roads. And on August 31, 1903, even the most skeptical agreed that Packard's "Old Pacific" had done something remarkable by arriving in New York after a fifty-two-day transcontinental journey from San Francisco.

In France, now, the Renault brothers had perfected a three-speed gear transmission. By 1899 one of their cars had completed a 3,600-kilometre tour through the Alps and Europe. By 1900 they had an engine of full 3½ horsepower with thermo-syphon cooling; Louis Renault had actually used it, to the astonishment of everyone, in army maneuvers. All over Europe the cars of Renault, of Daimler, and of Benz were racing in competitions marked by dust, smells, and disaster, and in the United States the automobile was becoming the adjunct of society.

All this while another idea was growing. It had taken root in the mind of one Rudolf Diesel when, as a student at the Munich Technical College, his friend Professor von Linde had impressed upon him the poor thermal efficiency of the steam engine. After his graduation in 1879 Diesel began to develop a modest engineering firm. He buckled down to the study of a new type of internal-combustion engine which, unlike those of Daimler and Benz, would burn only heavy crude oil and, more important still, would do away with impedimenta like hot tubes, open flames, or electrical sparks for ignition.

Diesel worked out most of his details mathematically before a single casting was machined. By 1893 his first engine was constructed and ready for test. Diesel turned it over and injected a slug of crude oil into the cylinder—whereat his engine rewarded him by blowing up viciously in his face. He was all but killed, but he grinned happily in his bandages, for he had proved his vital point: air could be so highly compressed that oil injected into its containing cylinder would explode from the heat of the compression.

With that point established, and himself recovered, Diesel soon designed a new machine—water-jacketed and with an improved air and fuel injection pump. His first reliable engine was constructed in 1897 at Augsburg, Germany. It developed 18 horsepower—in a year when the gasoline engine was still struggling in its rachitic infancy. Furthermore, of all prime movers, it came nearest to the ideal of the Carnot cycle.

In this year Adolphus Busch, King of American Brewers, was in Europe. On his way home he heard of Herr Diesel and his engine. A meeting was arranged in Cologne and Busch purchased the American patent rights. There then began an independent Diesel development in this country which, by 1906, was producing engines of 300 to 500 horsepower. Diesel fuel was cheap; there were few mechanical complications, and the engines demonstrated a gratifyingly high efficiency.

Yet it took a major war to bring to the Diesel engine the appreciation which it really deserved. It was a boon to the designers of submarines long before the war, but war focused world attention upon the means by which they were surface-driven. Thus it was that after the war Diesel engines came into favor not only for submarines but for many electric installations, particularly on ships. To-day, here and there, steamships are giving way to "Motor Vessels," with tremendous Diesel engines driving their main propeller shafts.

The inventor of this new machine never saw the more spectacular fruits of his victory over his first vicious engine. Following a lecture tour in the United States in 1912–13, he journeyed to England in September, 1913, and took the Harwich-Antwerp mail steamer *Dresden* to return to Germany. And while the *Dresden* trudged across the choppy waters of the English Channel, Herr Rudolf Diesel disappeared.

CHAPTER XXVII

Wanted—a New Watt

After almost two hundred years, the so-called machine civilization of so-called progressive countries still finds itself on the defensive, with the steam engine, which created it, periodically called upon to show whether, in the last analysis, it has been a blessing or a curse to the men who have inherited it. That is a question for philosophers—a race of men whose amusement it is, in the words of that salty cynic, Henry Adams, "to suggest unintelligible answers to insoluble problems."

Thomas Carlyle was perhaps the greatest and bitterest critic of the steam engine yet to have put a pen to paper. Writing in 1843, when England was still suffering from the worst dislocations of the industrial revolution, he painted a terrible picture of the smoke-stained and fume-withered nineteenth century in contrast with the clear, simple, and sunny twelfth. To-day the picture retains a certain terrible cogency.

Yet it does seem too bad to blame the unassuming steam engine for the follies and rapacities of man. It may be, as is sometimes claimed, that the steam engine has not lightened man's burdens but has instead only accelerated the pace of them. But if power has been responsible for an increase in duty, it has also been responsible for an increase in privilege. Perhaps the critics of power might be wiser, instead of sighing for its abolition, to discover how it might be further improved. It may be that a great many of the woes of the present day are all of a piece with the woes of Thomas Carlyle's nineteenth century and that the pains of modern industrial civilization are still growing pains—the discomforts of transition. For even to-day the most literal and unimaginative engineer has no great cause to be enthusiastic over the state of his art. Measured by the past it is gigantic; measured by a decent ambition for the future it is infinitesimal.

Regarded purely from an engineering point of view, the most tremendous power installation in the world is only the minutest reflection of solar energy—for that vast ball of flame, ninety-six million miles away, is the only source of power we now have, and through only two channels can we use it. Flick on the switch of your reading lamp and you begin the consumption of solar energy. But there is one thing you cannot tell about it—whether it was energy stored on the earth three weeks ago or 300 million years ago. Between these two ages of energy you have no option. If the generator which lights your lamp is driven by falling water, then the sun stored its energy only a few weeks ago, when it evaporated the rains and raised them from lower to higher ground. If coal or oil drives the generator, then it was 300 million years ago that the sun stored its energy by baking the dead forests into layers of carboniferous material.

Now the efficiency with which all man-made engines for the release of solar energy operate is a very melancholy matter. Mechanical efficiencies are good—which is to say that a very small percentage of power is lost at bearings and in other ways through unintentional reconversion into heat. The limitation lies deeper than that.

The steam engine in particular is a wastrel. Mechanically it does well enough; but the nature of water is at fault. It was the French physicist Carnot who some years ago showed, with a chart that made engineers of the nineteenth century silent and sad for a long while, that the steam boiler-engine combination was of necessity a comparatively inefficient one and could never be greatly improved. With such a machine, Carnot showed, one pound of coal could deliver just so much and no more energy. Everything, when the last word was said, depended upon the range of temperature between boiler and condenser. Much of the heat from the fuel escaped up the stack with the smoke; it was not practical to keep the condenser at a temperature lower than that of the room. And even with all the coddlings of forced draft there was a limit beyond which coal could not heat water without excessive pressures. But that need be no brake upon ambition to discover some entirely new system for the utilization of solar energy.

How poorly we have succeeded, so far, was made evident as early as 1882 by the engineer, Buchanan, whose calculations showed him that the sun, shining upon a terrain which put no meteorological impediments in his way, beat down upon the earth with energy amounting to 3,245 British Thermal Units per square foot per minute. This was in Egypt, one of the hot places of the world. The figures will become more spectacular when it is realized that they are the equivalent of 214,000,000 horsepower to every square mile of such territory. In other words, the sun in Egypt supplies 160 times as much power per square foot as would be represented if the same ground were covered with horses.

Engineers may well feel dissatisfied that, in the first place, their engines utilize too insignificant a fraction of nature's total daily gift, and that, in the second place, they waste so unconscionably much out of every fraction which they do develop. The sun being the source of all earth's energy, it is not strange that for some time engineers have sought some way of eliminating fuels like coal and oil and converting solar energy more directly. For the elimination of fuel many schemes, mostly through the use of mirrors, have been tried to concentrate solar energy upon one spot. One of the first of such attempts was made by the gifted John Ericsson who, even in his own time, was afraid that the world's coal supply would soon (that is to say, in about two thousand years, which is soon to any forward-looking mind) run out. His experiments showed that the rays of the sun in New York state falling upon a surface of ten square feet would evaporate 489 cubic inches of water per hour and would provide

> . . . sufficient heat to develop a force capable of lifting thirty-five thousand pounds one foot high in a minute, thus exceeding one horse power.

Had he worked in Egypt or in Arizona his results would have been still better.

> The field [Ericsson then went on to say] therefore awaiting the application of the solar engine is almost beyond computation, while the source of its power is boundless.

Ericsson worked on this idea for some twenty years, but his first sun engine was not constructed until 1870. Five years later he built seven of them, all consisting of a light frame lined with flat mirrors set on a curve to converge heat against a boiler. The results were not startling, but Ericsson continued to be fascinated by the idea. He now claimed 100 square feet of surface could be made to produce 1,850,000 foot-pounds of energy per hour and said further:

> We learn that 22,300,000 solar engines, each of 100 horse power, could be kept in constant operation, nine hours a day, by utilizing only that heat which is now wasted on the assumed small fraction of land extending along some of the water fronts of the sunburned regions of the earth.

Following Ericsson there have been many attempts to tap this same vein of energy, and some sun engines are operating to-day, not only in Egypt but in the United States. Most of them use concave mirrors or lenses to concentrate sunlight upon a steam boiler. They work, but not economically. The problem still waits for solution, just as Newcomen's engine waited for Watt to make his simple answer, "Use steam for pressure, not for vacuum."

Experimenters with the sun engine belong to that class of engineers which is satisfied with water but not with coal. Those who, on the other hand, are satisfied with fuels but not with water as a vehicle for storing heat likewise have their problems. To this group, also, John Ericsson belonged. In the hope of superseding steam, Ericsson built one of the first of his "caloric engines" in London in 1833. Its principle was that fuel, instead of heating water into steam, would make use of the expansive qualities of heated air.

This was not new. But not until Ericsson arrived upon this thermodynamic battleground did the idea really begin to develop. Ericsson spent thousands of dollars and vast energy in building his caloric engines after 1833 when his first experimental design of five horsepower seemed a success. In a small way the caloric engine prospered and Ericsson sold a thousand of them within two years. They were used for water pumps attached to hoisting gear. The Fitchburg Railway in Massachusetts owned one, and boasted proudly that it pumped 1,600,000 gallons of water at a total cost of $25 for fuel and oil and with no attention from an engineer. The New York Central Railroad had no less than forty-eight, and asserted that one engine at eleven cents per day did the work of five men. At the height of its glory the caloric engine drove the presses of no less than forty newspapers, including the New York *Evening Post* and the Hartford *Times*. The wags of the day had it that Mr. Ericsson in selling a hot-air engine to a newspaper was carrying coals to Newcastle; but the merits of his machine, its economy, lack of necessity for constant supervision, and absence of troublesome feed water rose lightly over these pleasantries. An attempt was even made to substitute caloric engines for the horse cars in the streets of New York, but here the Conservative Party rose to its full height and denounced the impiety. It was unnecessary. The thermodynamic difficulties in the way of the caloric engines have never yet been removed, and the greatest pride of John Ericsson's life is reduced to-day to the status of a Christmas toy.

All heat engines of whatever sort depend for their efficiencies upon the temperature ranges over which they operate. Some engineers dream of projects to create this temperature difference without conventional fuels by tapping deep shafts into the earth, whose temperature two hundred miles below the surface is over 500 degrees Fahrenheit. Others would sink a shaft many feet into the ocean and create the temperature difference by using unusually cold temperatures at the exit end of the engine instead of abnormally hot ones at the entrance. Still another group, more conservative and with an eye for immediate practicalities, would maintain the use of a conventional fuel but would substitute a new fluid to replace water in an engine's boiler.

Water is far from the ideal fluid for engine operation. True, it can be superheated enormously with consequent great efficiencies but superheating means high pressures. We are apt to think that, because a certain relationship of temperature and pressure is unavoidable in dealing with water, pressure is the desideratum. Strictly speaking this is not so; the desideratum is temperature range. If, then, a handy fluid were available which would acquire a temperature closer to that of the gases going up the power-house stack without soaring off into excessively high pressures, it would obviously be the part of thermodynamic economy to make use of it. Such fluids do exist. The only reason that Thomas Savery and Thomas Newcomen and James Watt thought of extracting energy from overheated water was that water was the limitless and all-pervading fluid with which they were familiar. For more than two hundred years the idea of water, and water alone, limited the thinking of almost everyone. Then, in 1912, along came W. L. R. Emmet, of the General Electric Company, who thought, speaking of liquids, What about mercury?

Water boils at 212 degrees Fahrenheit. Mercury boils at 677 degrees Fahrenheit. It is quite obvious, then, that mercury can absorb a great deal more heat from fuel than can water before this heat begins to spill over in the form of vapor at high pressure. For five years Emmet experimented with the idea of mercury as a boiler fluid, and by 1917 he had evolved a system full of promise—so full that in 1923 the constantly progressive South Meadow Power House at Hartford, Connecticut, made an experimental installation of a mercury-vapor engine. It cost $500,000 to install and represented Emmet's fifteenth attempt. The simple principle upon which it operates is in odd contrast with the bewildering complications of its practical engineering.

The fuel is ordinary fuel but it burns under a special boiler. This boiler contains mercury which, when heated slightly above its boiling point of 677 degrees Fahrenheit, passes over to a single-stage turbine at 35 pounds pressure. Having delivered its work by driving the rotor, it emerges now into its condenser at 455 degrees, once again liquid, but sufficiently hot to vaporize water into steam at 200 pounds per square inch. The steam thus generated drives another turbine. The combined result of these two operations delivers 52 per cent more power from every pound of coal than any water system hitherto devised. This is unprecedented. But mercury has these two disadvantages: it costs $1 a pound and in its vapor state is intensely poisonous. To offset these, there is a further advantage, which is that the mercury is capable of complete condensation and therefore of complete recircularization, being pumped back from the condenser into the boiler in a closed cycle.

So successful is this development so far that the General Electric Company is building for itself a $4,000,000 plant to drive its works at Schenectady—included in which will be the largest mercury-vapor turbine ever built. The capacity of the turbine will be 26,700 horsepower, and the mercury required in the boiler will weigh 250,000 pounds.

But mercury is not the only fluid which might replace water. Sulphur dioxide was used in Germany before the war —exhaust steam being used to vaporize it for the purpose of driving another engine. To-day an organic chemical known as diphenyl oxide seems a promising candidate. It lacks the disadvantages of mercury and into the bargain may produce efficiencies even higher than those of the mercury turbine.

There are likewise engineers who would like to dispense, in their prime movers, with fuels and boilers both. A perennial suggestion for the accomplishment of this lies in the idea of harnessing tidal drops, the most notable being the fifty- to seventy-foot drop in the Bay of Fundy. It is there that most of these experimenters turn their eyes. One such project for tide utilization is the construction of dams to receive waters at flooding tides, then keep them pent until ebb, releasing the energy then through water turbines. Some system such as storage batteries would of course be necessary to act as a reservoir of electrical current during the time that the generators were idle.

Let it not be forgotten that water has other forms than liquid and vapor. When water solidifies in a closed container the ice that forms exerts a pressure of 13½ tons to every square inch. It is, quite literally, a force to move mountains—for the erosion of rocks is in large measure brought about by splitting from frost. At present it is difficult to imagine a useful and continuous engine which would utilize this enormous pressure—but there it is, and perhaps only a new Watt is needed to show the way.

One hundred and fifty years ago steam power had only one use—that of pumping water. Slightly later it added to its stature by driving mills and machinery. Still later it was applied to the propulsion of land vehicles and then to propelling vessels at sea. With Hero of Alexandria, the three parts, boiler, cylinder, and pump were practically one. The Marquis of Worcester separated boiler and pump. Thomas Savery added the virtue of rapid condensation, and the engine, as a simple machine, was now perfected. With Newcomen the machine began to split up. Boiler and cylinder were differentiated, and by the time of Watt the condenser split off as one more organism, while the piston became double-acting and the higher pressures, plus the cut-off valve, added enormously to efficiency. The improvements which followed Watt were improvements in function of old parts, not the creation of new.

What will happen next no one can rightly tell. The words of Elijah Galloway, uttered almost one hundred years ago, still ring true:

> But, whatever may be our opinion of what has hitherto been done toward the success of this noble project [the railroad], we have no right to form any conclusion, as to what may be hereafter effected. It is well known that greater difficulties than any of these have been overcome, by talent and perseverance; and there is no reason to doubt, but that other Newcomens, and Smeatons, and Watts will arise, to vanquish those which remain.

CHAPTER XXVIII

Peak Load

It began so short a time ago, this age of power, and from such simple things. The Marquis of Worcester, musing over his cannon, burst by the force of overheated water; Sir William Gilbert, rubbing glass rods and amber and puzzling over their attractive power—in many ways these two men laid the paths which led to steam and electricity. Down one path have walked Savery, Newcomen, Watt, and Parsons—down the other Volta, Oersted, Ampère, and Faraday. When, at the close of the last century, these two paths converged and these men joined the forces of steam and electricity, progress completed a cycle in our modern eyes. Certainly vast steam turbines turning equally vast electrical alternators have spread a broad road down which our modern feet may tread.

Despite to-day's enormous outpouring of power and our utter dependence upon it, the generation and transmission of electrical current is still a rather haphazard business. For all its technical wonders it remains in many ways wasteful and random in its economics. Everywhere the tendency of the age is toward coalescence and concentration. Everywhere production and distribution operations have been gathering themselves together into fewer and fewer organizations of huger and huger mass. Upon the threshold of just such a development power stands to-day. The newest proposal goes by the name of giant power.

During the 1880's the transmission of electrical power was so uncertain and inefficient a business that Edison's first commercial station could serve an area of only one square mile. Even by 1890 this efficiency was not much improved. So modest were the capabilities of electrical transmission then that engineers, finding it necessary to develop power from a waterfall and utilize it three miles away, were still driven to connect their hydraulic turbine to an air compressor and transmit power over this distance by compressed air. In that year this was the longest transmission line in the world; ten years later, at the turn of the century, electricity could not economically be transmitted more than twenty miles.

But the development of alternating-current machinery and the transformer soon began to alter matters. The electrical generation of alternating current at Niagara Falls was the first great step which conquered electricity's weaknesses and gave it a rank and penetration which the new world needed. To-day a single power house can have a radius of some three hundred miles for the transmission of its energy.

Hydraulic power seemed at first ideally adapted to this type of high-tension generation. Indeed it is, with this one unhappy exception—there is not enough of it that can profitably be exploited. It was Charles P. Steinmetz who made the calculation that

> . . . if all the potential water powers of the country were now developed, and every raindrop used, it would not support our present energy demand.

Thus it is that less than 20 per cent of the electrical power generated in the United States comes from falling waters. Nor is this percentage likely to change unless the near exhaustion of coal and oil so greatly raises their costs that in desperation engineers will be forced to develop hydraulic power sources which at the moment are not economically worth while. Yet even then the eastern United States cannot greatly profit without even better transmission—70 per cent of the country's potential water power lies in, or west of, the Rocky Mountains.

And so, even after Niagara, steam boilers heated by coal continued to supply the bulk of the country's energy and steam locomotives continued to haul long twisting trains of coal cars up gradients, around curves, into cities and out of them again to carry the major source of power from the mine mouth to generating stations. This went on for a long time before a quizzical thought began to rise in the engineering mind. It was to the effect that the modern world had not devised this idea but inherited it. If it were sensible to use locomotives to haul more coal for hundreds of miles before power was generated from it, then it would be equally sensible by the same reasoning to bottle up the waters of Niagara in tank cars, haul them for five hundred miles to New York City, hoist them several hundred feet in the air, and there pour their contents over water wheels. This procedure would not be much more absurd than is the present one of carting coal back and forth thousands of miles every day all over the eastern United States. The more the engineering mind thought of this the more it became convinced that while engine and generator efficiencies were being vastly improved, the basic economics of power generation were no more than a blind inheritance from the eighteenth century.

It was not until the United States entered the World War in 1917 that the first step in the direction of giant power was taken. The federal government, which in peace times still takes delight in breaking up economic consolidations, in wartime changed its tune and followed obvious common sense by commanding river power stations in western Pennsylvania and eastern Ohio to interconnect and pool their flow of power for the benefit of the country's hard-pressed steel mills. To-day interconnecting stations can supply power over distances as great as fifteen hundred miles.

But that is not the whole story. Before New York City stations can generate power for light and traction, thousands of tons of coal must be hauled from the mine mouths at Pennsylvania and unloaded into bins, and in the transportation of one ton of coal for one hundred miles some twelve pounds of it must be burned up in locomotive fuel. The obvious answer to this economic waste is to give up this unnecessary transportation activity and erect great generating stations at the mine mouth. Then, instead of transmitting energy by coal cars, let us burn this same coal at the mine and with it generate electricity, which so much more easily, efficiently, and economically can be distributed wherever it is needed. The idea is not only plausible—it is sensible and inevitable. Some day the construction of such generating stations and the interconnections of many generating plants on the same transmission network will effect as great a change in the world of power as it saw in the days of James Watt and of Michael Faraday.

Beyond that point, prediction has little value. The complexity of the power plant idea may raise an occasional qualm—the world has already seen several times that as soon as any one system reaches a perfection beyond which ambition can think of nothing more, that system then begins to be supplanted by something else. The coach makers had reached a high state of their delicate art at just about the time the steam engine really relegated the stagecoach to oblivion. The creation of the sailing vessel had come close to perfection in the American clipper ship just at the time when the steamboat was ready to annihilate it. So it can

General Electric Company

MERCURY TURBINE

PERHAPS the forerunner of a new era in power generation

United Electric Light & Power Company

MODERN COLOSSUS

A 214,000 horsepower turbo-alternator installed at the Hell Gate station, New York City

Niagara Hudson Power Corp.

POWER FROM WATER

Hydro-electric equipment at Niagara Falls

perfectly well be that conventional electrical transmission by wires has no sooner reached its finest flowering than the wireless transmission of power may sweep it into the discard and do for transmission lines exactly what the railroad did for the elaborate system of canals projected in the first half of the nineteenth century.

Even as long ago as 1890 Nikola Tesla had convinced J. Pierpont Morgan, the elder, that wireless transmission of power was something out of which money might be made. His promise has been long in its fulfillment. Yet certainly the idea is no more fantastic than the idea of Marconi, Fleming, De Forest, and others who believed that a man whispering to himself in Newark, New Jersey, might be heard perfectly in every sitting room in the United States without the aid of wires, or that New York and London would converse with each other by telephone across three thousand miles of north Atlantic space.

Let us return for a moment to realities. To-day the engineer's ability to compress tremendous units of power into small space is incredible enough. In the Seventy-fourth Street station of the Interborough Rapid Transit Company of New York remains a fascinating study in contrasts. Here is generated much of the power utilized by the subway and elevated lines of New York City. The equipment consists of a 100,000-horsepower Westinghouse triple compound turbine, three 40,000 Westinghouse double compound turbines, and three 7,500-horsepower Corliss compound reciprocating engines. The contrast lies between the old reciprocating Corliss engines, once the wonder of their day, and the new triple compound turbine. The triplex alone could drive a great part of the electric transportation lines in New York. At maximum load it utilizes 826,000 pounds of steam in every hour. But the most amazing thing about it is this: it is but 20 feet high and requires a floor space, without boilers of course (since the hay always requires more space than the horse), of only 52 by 50 feet. One of the old Corliss engines alone weighs the same as the entire triplex, and the triplex is more powerful than thirteen of the Corliss engines put together. It supplies the power of 100,000 horses in a space no larger than one floor of a moderate-sized dwelling.

In 1919 this triplex machine was the largest in the world. To-day its nose is out of joint. The record for size is now held by machines at the Hell Gate generating system of the United Electric Light & Power Company in New York. Each of two machines there has a capacity of 215,000 horsepower.

These are the machines which, regardless of whether we like it or not, are changing the lives of everyone. It is a change of constant acceleration. Julius Cæsar could have returned to the world of George Washington with very little confusion to himself; he would have found boundaries, costumes, and society very different, and would have bemoaned them, of course; but he would have been quite familiar with all the means to agriculture, transportation, and mining. But Ulysses S. Grant, who died in 1885, could not orient himself in to-day's world without the patient help of a psychiatrist.

Neither Cæsar, George Washington, Alexander the Great, nor Napoleon Bonaparte was able to effect, in the world in which he lived, one one-thousandth part of the change which began to come over it when the steam engine of James Watt first girded up its loins. It is quite useless to recount differences between these worlds—they immediately become only a quantitative table the significance of which escapes everyone. Is it important that a modern Minneapolis flour mill can turn out 17,000 barrels of flour a day whereas an Athenian miller could produce but two? Have we saved our souls because a workman in 1830 could make 30 needles per day whereas one to-day can make 500,000? Or because a revolutionary shoemaker could produce one pair of shoes in four days whereas a modern workman takes only one day to make 10? Or because we travel by train between Boston and New York in 4¾ hours whereas in 1780 the journey took 12 days?

This book did not set out to answer these questions, and on them it will keep silence to the last. One hundred years ago Thomas Davenport's wife wept when her husband tore her wedding dress to shreds to make insulation for the first useful motor. To-day the descendants of that motor, for the greater leisure of ladies, drive vacuum cleaners, fans, refrigerators, sewing machines, hair dryers, egg beaters, and dish washers—and ladies are still weeping.

It is all very puzzling. If the millennium which James Watt and Michael Faraday thought they were bringing into the world has not arrived, certainly that is not their fault nor the fault of the scientists and engineers who have followed in their footsteps. Rather it is the fault of the rest of the world, equipped as it is with politicians and sociologists who have not been able to readjust their own arts and sciences to keep pace with the tremendous onrush of mechanical improvements. Matthew Boulton was quite right upon that eighteenth-century day when he thumped the back of his chair and told James Boswell that in the steam engine he possessed something that all the world desired. If the world of to-day has changed its mind, then certainly Matthew Boulton and his followers have been, like Ophelia, the more deceived.

Let us, at least, correct one wrong impression. The men who brought power into the world are not properly described by the term "crass materialists" so often applied to them. Of the two great media of power, steam and electricity, one is imponderable and both are invisible. These are no materials; and there was little materialism in the minds that worked with them. What they sought to solve and to create, and what their successors to-day would equally solve and create, is something very different—a mystery, and a religion.

THE END

INDEX

INDEX